POPULAR MECHANICS

HOME HOW-TO

PLUMBING AND HEATING

ALBERT JACKSON AND DAVID DAY

HEARST BOOKS New York

Library of Congress Cataloging-in-Publication Data

Jackson, Albert, 1943–
Popular mechanics home how-to : plumbing and heating /
Albert Jackson and David Day.
p.   cm.
Includes bibliographical references and index.
ISBN 0-688-10407-X
1. Plumbing—Amateurs' manuals. 2. Dwellings—Heating and
ventilation—Amateurs' manuals. I. Day, David, 1944-
II. Title.
TH6124.J33   1992
696'.1—dc20   91-35102
                                                         CIP

Printed in the United States of America

First Edition

1 2 3 4 5 6 7 8 9 10

CONTENTS

Cross-references

When additional information about a subject is discussed in more than one section of the book, the subject is marked in the text with the symbol (>) and the cross-references to it are listed in the margin of the page. Those printed in bold type are directly related to the task at hand. Other references that will broaden your understanding of the subject are printed in lighter type.

SEE ALSO

◁ Details for:
Obtaining
permission 70

UNDERSTANDING THE SYSTEM

Today, more and more homeowners are willing to tackle their own plumbing repair and remodeling projects. As with most do-it-yourself endeavors, cost is a motivating factor. The lion's share of most plumbing bills is the cost of professional labor, so doing your own repairs and installations makes good economic sense, and offers as a bonus the satisfaction of greater self-sufficiency.

Fortunately for the do-it-yourselfer, the plumbing supply industry has wasted no time in recognizing the growing consumer market open to them. Instead of selling plumbing supplies in bulk quantities to wholesale houses and plumbers, manufacturers are now marketing for the consumer as well. They are also making their products easier to use. Repair and replacement parts are now likely to be

attractively packaged with instructions and a "you-can-do-it" pep talk on the back of each blister pack.

To the further benefit of the homeowner, design and manufacturing trends have also steered the industry in the direction of lighter, less expensive and easier to install materials. City code authorities have recognized the need to make plumbing more affordable in new construction and have adjusted their material requirements accordingly.

The upshot of all of this is that basic plumbing has never been easier and has never required so few specialized tools. Still, the most important ingredient in any do-it-yourself project is confidence. Confidence starts with a basic understanding of how plumbing works and increases with every project you complete.

PLUMBING REGULATIONS

If you own a single-family dwelling, you can work on any of its plumbing. Your work will have to meet accepted plumbing standards as defined by local code regulations. Plumbing codes are written and enforced locally in the United States, but are based on specifications of the national *Uniform Plumbing Code.*

Before starting any major plumbing project, check with your local code office to see if the work you have planned meets specific code requirements. Any major project involving new piping or piping changes will require a permit and a series of inspections. Most plumbing appliance installations also require permits.

Plumbing standards are important because they protect us from health-threatening plumbing materials and practices. They benefit us all and should be supported by everyone.

A typical plumbing system

A residential piping system can be divided according to function into five basic categories. These categories are: pressurized water and fuel pipes, gravity-flow soil pipes, vent pipes and fixtures and appliances. All pipes in a system serve specific fixtures (sinks, lavatories and tubs) and appliances (water heaters, disposers, dish washers and clothes washers).

Take time to familiarize yourself with this network of pipes in your home before beginning any repair or

remodeling project. An understanding of how each component in your system works with other components will help you feel more confident about the project at hand. Think about each pipe, valve and fixture in your home as it fits within one of the five basic categories of function. Consider each item within the system separately. Until each element is examined according to its use, the system as a whole will likely remain a mystery. Plumbing is better learned one step at a time.

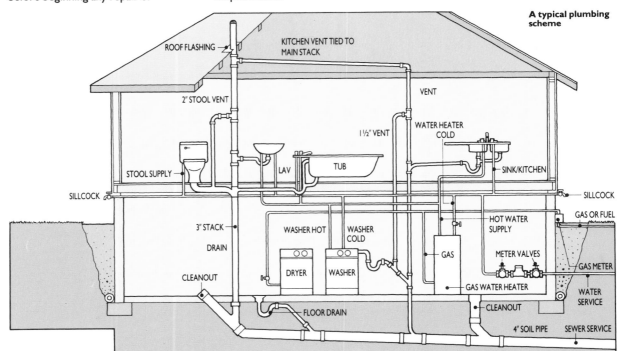

A typical plumbing scheme

A GENERAL OVERVIEW

A brief look at the fundamental components of a plumbing system will help you think through any problems you may be having.

Supply lines

Cold water service

The water that flows from your tap enters your home through a single water service pipe (**1**). This pipe is buried below frost level in your yard and extends from the city water main to a meter valve just inside your home. If water is supplied from a well, it flows to a pressure tank where it is stored until needed. In some cases, the water meter will be located in a meter pit just outside the home.

Cold water lines

The water service is then connected to a meter, which measures the amount of water you use. From the meter, a single trunk line usually travels up and along the center girder of a home. At convenient intervals along the way, smaller branch lines extend from the trunk line to service various fixtures throughout the house.

Hot water lines

At some centrally located point, a cold water trunk line branches off to enter a water heater tank. Some of the incoming water flows past the heater to other cold water outlets, but some enters the tank and fills it. The hot water outlet pipe extends from the other side of the tank and typically rises to the floor joist next to the center beam, becoming a hot water trunk line. This hot water trunk usually travels side-by-side with the cold water trunk and branches off to serve the hot water requirements of fixtures and appliances.

Drainage lines

Drainage lines are the pipes that carry soiled water from your fixtures to the public sewer system. They generally consist of one or more vertical stacks with horizontal drains attached. It is important that drainage lines are installed at just the right pitch. Too little fall will cause water and solids to drain sluggishly and may cause the line to clog. Too much fall, particularly on long runs, will cause the water to outrun the solids, which may also cause the line to clog. A fall of ⅛ inch per foot is ideal. Use a level on each pipe.

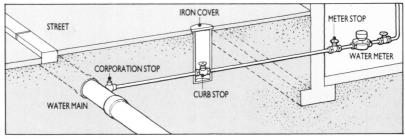

1 A typical water service

Fixture traps

As a public sewer system is vented through the roofs (via plumbing stacks) of the homes connected to it, sewer gas is always present in large quantities in drainage systems. Each fixture must therefore have a seal to keep noxious sewer gases from entering the home through fixture drains. This seal is created by a U-shaped pipe, called a trap, beneath each fixture. Traps allow water to pass through them when a fixture is drained. After a fixture is drained, however, the trap holds a measured amount of water in its bend so that sewer gas cannot pass through the pipe and into your home. Every fixture in your home must have a trap, and every trap must be vented.

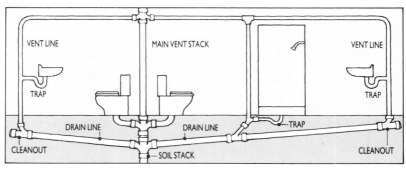

2 Stack-vented toilets with re-vented lavs and shower

Venting

Venting is one of the most critical aspects of a drainage system. Without properly vented drain lines, stools won't flush properly, sink and lavatory drains choke and high-volume appliance drains may overflow. But most importantly, unvented drain lines siphon water from fixture traps. Once a trap seal is broken, sewer gas quickly enters your living quarters. Sewer gases can cause respiratory problems and headaches in quantities too small to detect by smell.

A fixture trap may be vented in one of two ways. The simplest way is to size horizontal drains so that the top half of the drain serves as the vent for that fixture. This is called stack venting (**3**).

The second method is called *re-venting* (**2**). A re-vent is a separate vent that is installed on the highest part of a drain pipe and within a few feet of the trap. It then extends through the roof independently or ties back into the main vertical stack above the flood plane of the highest fixture, often in the attic. The type of vent used depends upon the number of fixtures and floor levels involved and the structural constraints peculiar to each home.

SEE ALSO

Details for: ▷
Emergency repairs 7
General piping
considerations 8

● **Gas and fuel lines**
Natural gas and fuel oil piping will often run next to the plumbing pipes in your home. Because appliances such as clothes dryers and water heaters are often gas-fired, you will need to learn enough about gas piping to service and replace these appliances.

Gas is brought to your meter through an underground service pipe made of plastic or coated black iron. Its pressure is reduced from around 25 psi to about 6 ounces before entering your private system. Once inside your home, it is piped directly to your gas appliances.

Fuel oil piping is usually plumbed in iron and makes its way to the furnace under low pressure gravity flow.

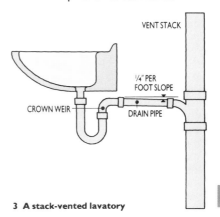

3 A stack-vented lavatory

DRAINING A WATER SYSTEM

Being able to locate and shut off your meter valve or, if you live in the country, your pump switch, is an important homeowner responsibility. You must be able to act quickly in an emergency.

Locating your meter

The main shutoff can be found at the water meter (◁). Water meters are made of brass and are usually not larger than 6 inches in diameter. At the top of the meter you will find a glass or plastic housing that contains a dial and a counter similar to the odometer on your car. Meters are often just inside the basement wall closest to the street. As previous owners may have covered your meter with cabinetry or a pipe chase, look for an access panel. Utility rooms and basement stairwells are also good places to look. If all else fails, follow the cold water inlet pipe from your water heater back to its source.

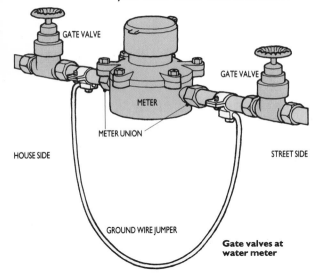

**Gate valves at
water meter**

Shutting off your meter

Once you have located your meter, notice that there is a valve on each side of it. Many meter valves have handles, but some will have only a bar that must be turned with a wrench. Both of these valves will stop water from entering your home. The valve on the street side of the meter will stop the water before it enters your meter. Conversely, the valve on the house side of the meter will stop the water from entering your plumbing system.

If you merely wish to shut the water off, as when you leave on vacation, or in the event of an emergency, either valve will do the job. If, however, you need to drain your water system, then both valves will need to be closed.

Draining the system

To drain your water system, first shut the street side of the meter off. Then shut the house side of the meter off. When both valves are closed, place a bucket under the meter and loosen the meter union next to the house side valve. The small amount of water trapped in the meter will trickle through the opened union. Then go through your home and open all faucets to prevent air lock. When all faucets are open, return to the basement and open the house side valve and slowly drain the water into the bucket.

Draining your water heater

There can be several reasons why you might want to drain your water heater (◁). The most likely occasion will be when removing an old heater. Sediment, faulty electrical elements and stuck relief valves are three other common problems that require draining.

Start by shutting off the water supply to the heater. If your heater has a cold-water inlet valve, use it and leave the cold water side of the system on. If no inlet valve is present, you will have to shut the water supply to the entire house down at the meter.

Once water to the heater is shut off, open all hot water faucets in the house to prevent air lock. Then, open the spigot or drain valve at the base of the heater and drain off as much as you need for the repair. If yours is an electric heater, one added precaution is in order: Because energized electrical elements burn out in a matter of seconds when not immersed in water, you will need to shut off the power to the heater *before* draining it.

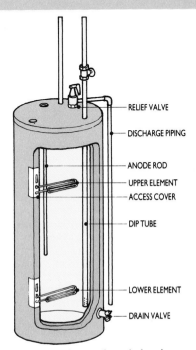

RELIEF VALVE

DISCHARGE PIPING

ANODE ROD

UPPER ELEMENT

ACCESS COVER

DIP TUBE

LOWER ELEMENT

DRAIN VALVE

Drain water heaters from drain valve

Partial drain downs

In many instances, such as toilet repairs, there will be no need to drain the entire system. A toilet usually has a shutoff valve between the riser and the supply line under the tank. Some homes have shutoff valves under sinks and lavatories as well. When these valves are present, use them. It is almost always easier to isolate a single fixture than to put the entire system out of order.

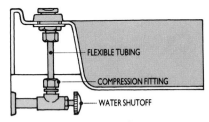

FLEXIBLE TUBING

COMPRESSION FITTING

WATER SHUTOFF

Independent shutoff under fixture

Recharging the system

Just as you opened all faucets to prevent an air lock when draining the system, you will need to open them to bleed air from the lines when recharging the system with water. Open the meter valves only part way. Then bleed the air from the newly charged lines. After all air in the system has escaped through the faucets, turn the faucets off and turn the meter valves all the way open. Small bursts of air may still escape through your faucets when you first use them after recharging, but all air should be dissipated after the first full pressure draw. If you do not bleed trapped air from supply lines, the shock of air released under full pressure could damage faucet and supply tube seals.

It seems that plumbing leaks occur when we are least prepared to deal with them. In such cases, a plumber may not be available, and you may not have the time or materials to make a permanent repair. There are, however, a number of stopgap measures that you can use to repair leaks temporarily.

A LAST-DITCH REPAIR

If conventional repair materials are not available, you can sometimes make do with materials found around the house or at your local all-night service station. To create a makeshift sleeve coupling, use a piece of bicycle or car tire inner tube and a few radiator hose clamps. Wrap the inner tube around the split pipe several times and clamp it in place with hose clamps. This is at best a stopgap measure, but it will slow the leak until you can make a more permanent repair.

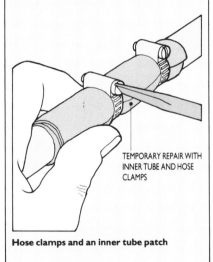

TEMPORARY REPAIR WITH INNER TUBE AND HOSE CLAMPS

Hose clamps and an inner tube patch

Frozen and split pipes

In colder climates, pipes located in exterior walls, crawlspaces and attics are often subject to freezing. The best preventive measure is to insulate these pipes. Even insulated pipes, however, can freeze when exposed to cold drafts of air. When a pipe freezes, a plug of ice forms in a small section of the pipe and expands against the pipe walls. The expansion swells the pipe, and in most cases, ruptures the pipe wall.

Even a well protected pipe may crack after years of use. Factory defects and corrosion are often responsible for these leaks. Regardless of how a pipe cracks or splits, emergency repair methods are usually the same.

A sure sign that a pipe has frozen is when no water passes through the pipe to the faucets nearest the freeze. You will often be able to feel along the pipe and locate the frozen area. If the pipe has not yet ruptured, use a portable hair dryer to warm the frozen area until water again flows to the nearest faucet. Once thawed, you should wrap the pipe

with insulation. If no insulation is available, use old rags and fasten them to the pipe with tape, string or wire.

If you can see that the pipe has already split, you will need to drain the system before thawing the frozen area. Once thawed, you will need to make some sort of emergency repair. If you are able to find a plumbing supply outlet, the best solution is to buy a sleeve-type repair coupling. These couplings can be purchased in a number of standard sizes. They consist of two metal halves that are hinged on one side and bolted together on the other. A rubber sleeve fits inside and wraps completely around the pipe.

To install a sleeve repair coupling, first clean the pipe with a wire brush or sandpaper. Then fit the rubber sleeve around the pipe so that the seam is opposite the leak. Fit the metal halves of the collar over the sleeve and tighten the two halves together. While this method is an emergency repair, it is also permanent.

SEE ALSO

Details for: ▷	
Plastic pipe fittings	9–10
Steel pipe	11
Repair fittings	12

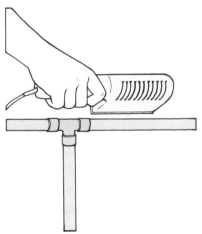

A hair dryer works well in thawing pipes

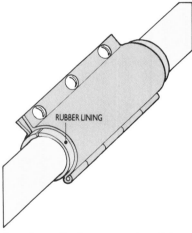

RUBBER LINING

A repair coupling for permanent repairs

Epoxy patch repair

If you do not have access to plumbing materials, you may still be able to buy a general-purpose epoxy kit. These kits consist of two sticks of putty that you knead together. Once the parts are completely mixed, you will have about fifteen minutes to work before the mixture begins to set. Start by cleaning and drying the area around the split with sandpaper and alcohol. Knead the two components until they are a single consistent color and begin to give off heat. Then press the epoxy putty

around the pipe. Smooth the ends with water until the epoxy forms a seamless bond around the pipe several inches on either side of the split.

Epoxy takes a full 24 hours to cure, but after a few hours, you should be able to turn the water on slightly. Do not put full pressure in the pipes for at least 24 hours after applying an epoxy patch. You will have to put up with slow-running faucets and toilets, but you will have the water you need to run a normal household until the epoxy cures.

EPOXY PATCH

Plumber's epoxy putty bonds to pipe

WORKING WITH PIPE

Until you learn to work with pipe, your plumbing capabilities will be limited to simple maintenance. While Neoprene gaskets, no-hub couplings and plastic pipe and fittings have greatly reduced the need for special knowledge and specialized tools, most remodeling still requires that you understand how traditional plumbing materials are put together. In most cases, these newer, easier-to-use materials will still have to be tied into existing pipes. You may also need to dismantle some existing piping in order to repair or extend your

plumbing system. In short, knowing how to use plastic pipe is of little use if you do not also know how to tie it to other kinds of pipe in your existing system.

The good news is that the skill needed to work with cast-iron, steel and copper pipe has been seriously overrated. You can do it. You may need a few specialized tools, but you can rent those. What you need most is a basic understanding of how these materials are put together and what fittings and tools make the job easier. The rest is a matter of practice.

Making the connection

With the acceptance of plastic as a plumbing material, several new methods of joining plastic to conventional soil pipe have also been developed. These connectors have virtually eliminated the skill once needed to form mechanical joints. Where once molten lead and oakum (an oily, ropelike material) were packed into bell and spigot joints, now Neoprene gaskets make the seal. Now, instead of dismantling a run of pipe all the way back to its nearest hub, you can join two hubless pipes in minutes with no-hub couplings.

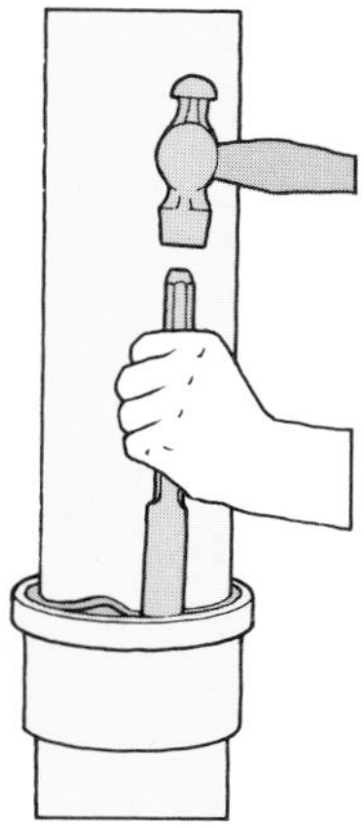

Bell and spigot gaskets

Neoprene gaskets are made for every standard size cast-iron soil pipe. They are fitted rubber collars that snap into the bell of a cast-iron pipe. The inside of the gasket is then lubricated with detergent (dish soap works well), and the male end of the adjoining pipe is forced into the gasket until it seats.

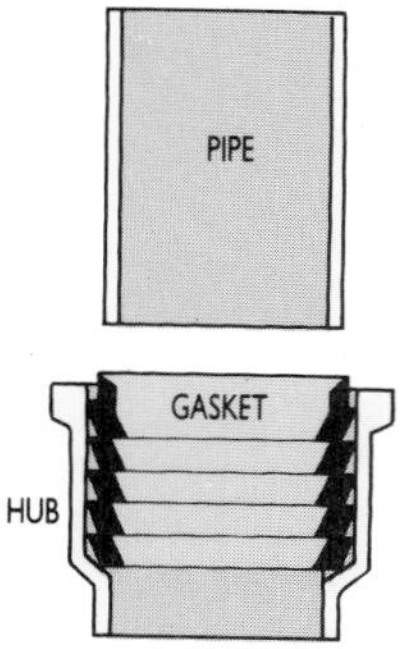

No-hub couplings

No-hub couplings are rot-resistant rubber sleeves with stainless-steel bands around them. They are designed for use on drainage pipes and are approved by most code authorities. To install no-hub

couplings, you simply slide one end of each pipe into the sleeve and tighten the bands. No-hubs come in many sizes, including increasing and decreasing couplings that allow you to join different pipe sizes to one another. They are particularly handy in joining dissimilar pipe materials, such as cast iron and plastic or plastic and copper.

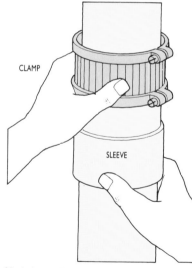

CLAMP

SLEEVE

No-hub coupling

Lead-and-oakum joint
Made by packing a hub with ⅔ oakum and ⅓ lead.

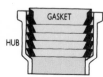

Neoprene gaskets
Neoprene gaskets are easy substitutes for lead-and-oakum joints.

DRAIN AND VENT SIZING CHART

FIXTURE	DRAIN SIZE	VENT SIZE	MAX. LENGTH STACK VENT
Toilet	4"	2"	10'
Toilet	3"	2"	6'
Sink	1½"	1½"	3.5'
Lavatory	1½"	1½"	3.5'
Tub	1½"	1½"	3.5'
Shower	2"	1½"	5'
Laundry	2"	1½"	5'
Floor drain*	2"	1½"	5'

* Vents waived by some codes.

GENERAL PIPING CONSIDERATIONS

When installing drainage pipe, there are a few rules you will need to follow in order to ensure mechanically sound joints and even flow patterns. They are as follows:

1. The grade, or elevation of a drainage pipe should not be less than 1/16 inch or not greater than 1/4 inch per running foot of pipe. In total, the fall of a given pipe run should not be greater than the diameter of the pipe involved. For example, a 2-inch pipe should not drop more than 2 inches along its entire length. If structural constraints require that a pipe drop more than its diameter, fittings should be used to "step" the pipe down to a lower plane. In this case, a re-vent will be required before stepping down.

2. When drainage pipes are wet, or carry water, fittings with gradual flow patterns should be used. For example, tees should be used only when they are the highest fittings on a vertical pipe. They should never be used in horizontal positions or when other fixtures are served above them. Because wyes offer much more gradual flow patterns, they should always be used instead of tees, except as the highest branch fitting on a vertical stack.

3. When drainage pipes are suspended from floor or ceiling joists, they should be supported by pipe hangers at a rate of at least one every 6 feet, or one for every pipe less than 6 feet long.

4. Every vertical stack must have a cleanout fitting at its base before entering a concrete floor.

5. All pipes installed underground or under concrete must be laid on even, solid soil. No voids or low spots are permitted under a pipe. If voids exist, or if the grade has been overexcavated, the ditch should be lined with fill sand. You should never fill voids or raise a pipe with soil. Soil is sure to settle with time, causing the pipe to sag and clog, or to shear off entirely.

When excavating a ditch for soil pipe, dig a small impression in the soil for each pipe hub. This will keep the entire length of the pipe from resting only on the hubs. Because soil pipe is buried permanently, either underground or under concrete, always work for a permanent installation.

PLASTIC DRAINAGE PIPE AND FITTINGS

Plastic pipe is the easiest pipe to handle because it is lightweight, can be cut with a hacksaw and is joined to its fittings with glue. Plastic drainage pipe comes in two forms. ABS pipe is black and PVC is white. Both are schedule #40 weight, which is the wall thickness required by code for *drainage pipe. There is no appreciable difference between the two, except that ABS has become more expensive in recent years and the plumbing industry in general seems to be moving away from it. In the interest of consistency, you should match the type already in your home.*

Cutting and assembling plastic pipe

To cut and fit plastic pipe, measure for the desired length and mark the pipe with a pencil. Then, using a hacksaw, cut carefully across the pipe (**1**). Take particular care in making straight cuts. A crooked cut will keep the end of the pipe from fitting properly into the bell of the fitting. Smooth the cut edge with a file (**2**).

It is always a good idea to assemble pipe and fittings before gluing to make sure that all measurements are accurate and all fitting angles correspond. Then, before dismantling, mark the fitting and pipe (**3**) at each joint so that you will have an easy reference point when gluing them together permanently (**4**).

Gluing plastic pipe

Gluing pipe and hubs is easy but requires accuracy and speed. Apply glue to the inside of the hub of the fitting. Then glue the outside of the pipe, covering a depth consistent with the depth of the hub (**5**). Press the fitting onto the pipe with the pencil marks about an inch apart. When the pipe is in all the way, turn the fitting so that the pencil marks line up. By turning the fitting on the pipe, the glue is spread out and the friction helps cement the joint.

Glued joints (technically, cemented

joints) set in about 30 seconds, so if you make a mistake, you have to pull the joint apart very quickly. Plastic pipe cement does not really glue one surface to another; rather, it melts the two surfaces, causing them to fuse. Once a joint has set, it is permanent.

When buying materials, be sure to choose a cement made for the type of pipe you buy. ABS glue will not cement PVC pipe and fittings. PVC glue does work with ABS pipe, so it is always better to use compatible materials.

Installing drainage pipe

Drainage pipe must often be installed in interior walls. Because plastic pipe expands and contracts with hot and cold water, make sure that holes drilled in the wall studs are large enough to allow for this expansion. Stud and joist holes should be at least 1/8 inch larger than the exterior diameter of the pipe. Plastic pipe should never be shimmed tightly against wood. Without room for expansion, plastic pipe will produce an

annoying ticking sound in the wall after hot water has been drained through.

Plastic drainage pipe can be joined to pipes made of steel or copper in two ways. One way is to use a male or female adapter. One end of the adapter is glued to the plastic pipe and the other end threads into a female fitting. No-hub couplings make quick and easy connections and are also good fittings for drain cleanouts.

Joining plastic to cast iron

When joining plastic to a cast-iron hub, use a Neoprene gasket. Fold the gasket into the hub and lubricate it. Then file down the sharp edge on the end of the plastic pipe and push the pipe into the gasket. To join plastic pipe to the hubless end of a pipe, use a no-hub (or banded) coupling.

Neoprene gasket for cast-iron hub

1 Tape paper around the pipe and cut

PAPER GUIDE

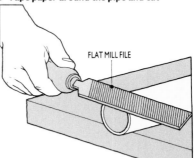

FLAT MILL FILE

2 File away any burr left by the saw

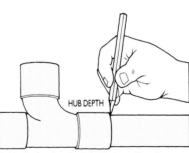

HUB DEPTH

3 Mark the depth of the hub on the pipe

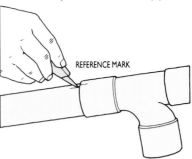

REFERENCE MARK

4 Mark pipe and hub as reference

5 Apply cement to both pipe and fitting

SEE ALSO

Details for: ▷

Steel pipe	11
Repair fittings	12

● **Supporting pipe runs**
Plastic pipework should be supported with clips or saddles similar to those used for metal pipe, but because it is more flexible, you will have to space the clips closer together. Check with manufacturers' literature for exact dimensions. If you plan to surface-run flexible pipes, consider ducting or boxing-in because it's difficult to make a really neat installation.

PLASTIC WATER PIPE AND FITTINGS

With one exception, CPVC plastic water pipe is put together the same way as plastic drainage pipe. Plastic pipe has a shiny residue on its surface that should be either sanded or treated with primer solvent before cement is applied. A fine-grit sandpaper works well when sanding the ends of each pipe and inside of a fitting, but solvent is faster and more thorough. Simply brush the surfaces to be glued, wait a few seconds and wipe them clean with a soft cloth. Then brush the glue on the pipe and fitting and push them together with a slight twist.

JOINING CPVC PIPE TO DISSIMILAR MATERIALS

Plastic water pipe can also be joined to steel and copper pipe by means of plastic threaded adapters. Both male and female plastic adapters are available. One end of the plastic adapter is glued to the plastic pipe and the other threaded into a fitting or onto a pipe. When joining plastic water pipe to existing metal piping, wrap the male threads with plastic pipe joint sealant tape. Because plastic female adapters can expand when threaded onto male threads, a better choice is to use a plastic male adapter threaded into an iron or copper female adapter or fitting.

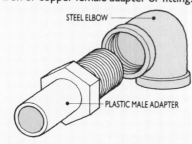

STEEL ELBOW

PLASTIC MALE ADAPTER

Adapters connect plastic to existing steel

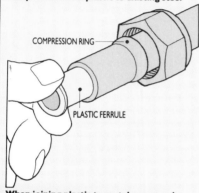

COMPRESSION RING

PLASTIC FERRULE

When joining plastic to metal compression fittings, use a plastic ferrule

Polybutylene pipe

PB pipe is a relative newcomer to the plumbing industry. It is not directly compatible with other plastics, but offers several real advantages. The first advantage is that it will take a freeze without splitting. The second is that PB fittings contain mechanical joints. They can be turned and adjusted in place and are therefore more forgiving of beginner error. And finally, PB pipe is bendable, which greatly reduces the need for elbow fittings. As PB piping is not universally accepted by code authorities, check to see if you can use it before starting a piping job.

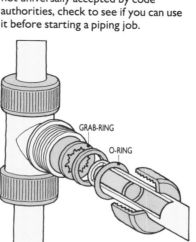

GRAB-RING

O-RING

Push-fit O-ring fitting
Just lubricate the pipe and push it in until it seats.

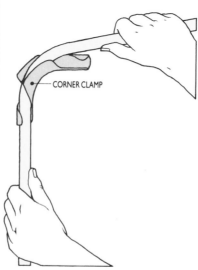

CORNER CLAMP

Bending polybutylene pipe
Polybutylene pipe saves fittings because it can be bent around corners.

The manufacturers of PB pipe have designed connector fittings to match nearly all water pipe materials and sizes. These fittings come in two styles. One is a simple compression fitting and the other has a push-fit O-ring combination. A good choice for supply pipe installations is the push-fit type. It goes together easily and almost never leaks. PB fittings adapt well to other materials, such as copper and CPVC. A threaded adapter is also available for threaded pipe hookups. PB valves come in many sizes for a variety of special-use situations.

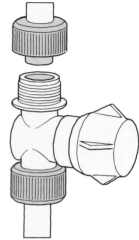

Compression valve
This valve is an ideal in-line toilet supply shutoff.

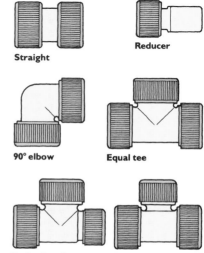

Straight

Reducer

90° elbow

Equal tee

End reduced tee

Branch reduced

Plastic pipe fittings
Some of the basic fittings for connections are shown.

Steel pipe is available in galvanized and black iron forms. Galvanized pipe is used for water supply lines and drainage and vent lines, while black iron is used primarily for gas piping. Though they are put together in exactly the same way, you should never mix the two. When galvanized iron is used on gas installations, the gas in the line will attack the zinc plating and cause it to flake off. These zinc flakes will be carried through the system and can easily clog the orifice and control valve of a water heater or furnace. On the other hand, if black iron pipe is used in water supply lines, it will rust shut in a matter of months.

Steel pipe is put together with threaded joints. Because steel was the predominant piping material of residential plumbing for the first 60 years of this century, you will find it in many older homes. You may never need to install steel piping, but chances are you will have to deal with it when making changes and repairs.

Steel pipe comes in 21-foot lengths and is threaded on each end at the factory. It also comes in short precut, prethreaded lengths called nipples that graduate in ½-inch increments from 1 foot to approximately 1 inch in length. When you need custom-cut lengths, you will have to cut and thread them yourself, using a die cutter.

Cutting and threading steel pipe

Steel pipe can be cut with a hacksaw. In fact, when cutting out a section of existing pipe, a hacksaw is your best choice. When you intend to thread the cut end, however, a wheel cutter will give you a much more uniform cut. Wheel cutters can be found at most rental stores along with the threading dies you may need.

To determine the exact length of pipe you need, measure between the two fittings and add the depth of the threads inside each fitting for your total length. Then mark the pipe with a soapstone or crayon. Place the pipe in a pipe vise and tighten the cutter on the pipe so that the wheel is directly on your mark. Then tighten the wheel one-half turn and rotate the cutter around the pipe. When the cutting wheel turns in its groove easily, tighten it another turn and rotate the cutter again. Repeat this process until the pipe is cut completely through.

When the cut is made, leave the pipe in the vice and get ready to cut new threads. Cover the first inch of the pipe with cutting oil and slide the cylinder of the die onto the pipe. Set the lock on the die to the "cut" position. Then, while pressing the die onto the pipe with the palm of your hand, crank the die handle. When the die teeth begin to cut into the metal, you will feel some resistance. You will then be ready to crank the handle steadily around the pipe to cut the threads (**1**).

About every two rounds, stop and pour oil through the die head and onto the new threads. This is very important. Without oil, the pipe will heat-up and swell until you can no longer turn the handle. Dry pipe threads will also ruin the die cutters in short order.

Continue cranking and oiling until the first of the new threads shows through the front of the die. Then, reverse the direction lock on the handle and spin the die off the pipe. Thread the other end and you will be ready to install that length. All threaded steel joints should be put together with plastic pipe joint sealant tape or pipe joint compound, on the male side only.

Plastic spacers
A dielectric union prevents corrosion by separating the metals with a plastic spacer.

Wire pipe hook

Pipe strap

Two-hole strap

Hole strap

Pipe supports
Use standard pipe hooks and fasteners to support iron pipes.

Pipe adapter   Washer   Pipe adapter   Plastic spacer   Fastening nut

IRON PIPE FITTINGS

You will find a fitting for nearly every possible piping configuration. Here are nine common fittings you will find at your local plumbing outlet.

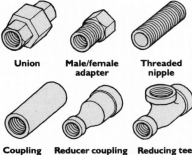

Union   Male/female adapter   Threaded nipple

Coupling   Reducer coupling   Reducing tee

90° ell   Reducing ell   Combination tee

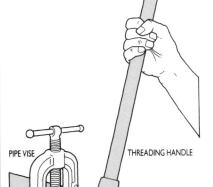

PIPE VISE   THREADING HANDLE

HEAD OF THREADER

1 Threading an iron pipe
Lock the pipe in a pipe vise and use a threading die.

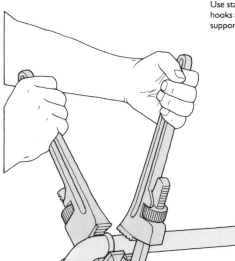

Removing a threaded fitting
Use a second pipe wrench to back-hold the pipe.

FITTINGS FOR REMODELING AND REPAIR

Repair and remodeling work often require that you cut a section of pipe from between two fittings. This is easily enough done with a hacksaw. Once the cut is made, you will need to back each section of remaining pipe from its fitting. Use two pipe wrenches, one to back the pipe out and one to hold the fitting so that no joints further down the line are disturbed. Otherwise, you might cause leaks.

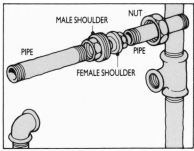
Fittings for reconnecting cast-iron pipe

Reconnecting with steel pipes, fittings and unions

To reconnect the removed portion of the line with steel, you will need a union. If a new tee or wye fitting is to be spliced into the new line, thread it onto a length of pipe and thread the pipe into one of the existing fittings. Thread another pipe into the other existing fitting. This pipe should come to within 6 inches of the new tee or wye. Finish with nipples and a union.

Reconnecting with copper or plastic pipe

Splicing in new pipe between two existing steel fittings is easier to do with copper or plastic. In each case, threaded adapters can be screwed into the steel fittings. In the case of copper, the remaining joints will be soldered. In the case of plastic drainage or water pipes, all other joints will be glued. PB pipe, of course, would be fitted with mechanical PB fittings and adapters. Because plastic water pipe is not universally accepted, you should check with your local code authorities before installing CPVC or PB piping materials.

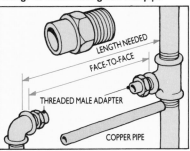
Fittings for splicing copper to iron pipe

Working with copper pipe

Copper pipe is available in four wall thicknesses. *Type K* is the thickest and is used primarily for underground water services and under concrete for supply lines. It comes in soft coils.

Type L offers the next thickest pipe wall. It is available in soft coils or rigid sticks. The soft version is typically used in gas pipe installations and is connected with flare fittings.

The thinnest allowable supply line pipe is *type M*. It is made only in 20-foot rigid lengths and is the most widely used in residential water systems.

The thinnest-walled copper pipe is *DWV*. As its initials imply, it is used as drain, waste and vent piping. Although seldom used now for drainage and vent piping, it was widely used in the '50s.

All copper pipe can be cut with a hacksaw, but if you intend to use flare fittings on soft copper, a wheel cutter (◁) will give a more uniform edge. All copper can be soldered, and all soft copper can be flared for use with flare fittings. Copper drainage pipe can be joined to steel, plastic or cast-iron pipe with no-hub couplings. Hard copper supply connections can also be joined with compression fittings.

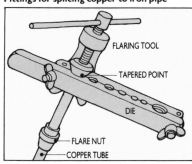
Flare fittings can be used with soft copper

Flare and compression fittings

As its name implies, a flare fitting requires that you flare the pipe to fit the fitting. Slide the flare nut onto soft copper pipe and clamp the flaring die within 1/8 inch of the end of the pipe. Thread the flaring tool against the pipe end until it expands evenly against the tapered die seat. Remove the tool and draw the fitting together with two wrenches.

A compression fitting operates in reverse fashion and so does not require a special tool. Simply slide the fitting nut and brass compression ring onto hard or soft copper pipe and thread the nut onto its fitting. Tighten hand tight, plus one to one and a half turns. Use pipe-joint compound on flare and compression fittings.

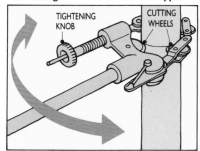
Cut cast-iron pipe with a snap cutter

Working with cast-iron pipe

You can cut cast-iron with a chisel by tapping lightly in a line around the pipe until it breaks. But the best way is to use a snap cutter. A snap cutter (◁) is a ratchetlike tool with a chain that wraps around the pipe and cutting wheels that clamp onto the pipe. Snap cutters are available at most tool rental outlets.

Traditional hub and spigot joints are now made with Neoprene gaskets. Where hubs are not available or where hubless cast-iron is joined to plastic, steel or copper DWV pipe, no-hub couplings make the easiest connection.

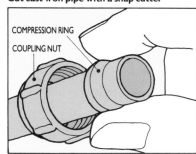
Compression fittings for copper-to-copper joints

Soap, hair, food particles and cooking grease all help to clog drainage lines. Occasionally a fixture trap will accumulate a blockage that can be forced clear with a plunger, compressed air or a blow bag. Most blockages build up inside pipes over an extended period of time, however, and must be cabled, or "snaked," to be opened. The method you choose will depend upon the fixture involved and the size of the drainage line.

Forcing fixture traps

Plungers, blow bags and cans of compressed air can all be used to free simple trap clogs. When forcing a clog from a trap with any of these, be sure to plug any connecting airways. When plunging a lavatory, for example, use a wet rag to plug the overflow hole in the basin. When forcing the trap of a two-compartment sink, plug the opposite drain. After the clog has been forced through the trap and into the drain line, run very hot water through the line to move the clog into the main stack or soil pipe.

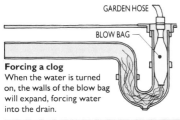

Forcing a clog
When the water is turned on, the walls of the blow bag will expand, forcing water into the drain.

GARDEN HOSE

BLOW BAG

Snaking a shower

If drain water backs up into a tub or shower from another fixture, it probably means that the main sewer line is clogged. If a tub or shower drains slowly or not at all, then you should snake the tub or shower trap and drain line. To snake a tub, remove the cover plate from the overflow valve and push the cable into the overflow pipe. Do not attempt to cable through the drain opening.

To clean a shower drain, start by removing the drain screen. If the screen is fastened to the drain with screws, remove them and pry the screen up with a knife. If the screen snaps in place, simply pry it up. Use a flashlight to look into the trap. If the trap is clogged by soap and hair, you can usually pick the clog out with a wire hook. If the trap is clear but the line is clogged, push the cable through the trap and cable the line to the main stack, or the nearest soil line.

Snaking fixture drains

When you wish to clean the drain line of any fixture with a snake, you will first have to remove the trap. Use a pipe wrench or adjustable pliers to loosen the nuts at the top of the trap and at the drain connection near the wall. With S-traps, loosen the nuts near the floor and at the trap weir. To avoid cracking or breaking a P-trap, hold it firmly and turn the nut with steady, even pressure.

With the trap removed, slide the snake cable into the drain line until you feel resistance. Then tighten the lock nut on the snake housing and slowly crank and push the cable deeper into the line. All bends in the line will offer some resistance, but slow, persistent pushing and cranking will force the cable around bends.

If the cable comes up against a blockage and begins to kink inside the line, back up a foot or so and try again. If you feel steady resistance, keep pushing and cranking. Every 3 feet or so, stop, pull the snake back a little and then crank forward again. Try to determine how many feet it is from the trap to the main stack and cable all the way to the stack, even if you break through the clog. When pulling the snake out for the final time, crank in a clockwise direction. If the clog was caused by a foreign object, like a piece of a sponge or dish cloth, the clockwise coil on the end of the cable will hang onto the object and pull it out as well.

After you have snaked the drain, replace the trap. If the trap washers seem hard and brittle, replace them with new washers. Because much of the debris from the clog will remain in the line, flush with plenty of hot water. If the fixture drain backs up, plunge it thoroughly and flush the line again.

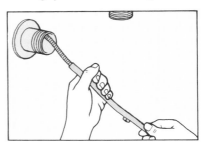

To clean a fixture drain, remove the trap

SEE ALSO

Details for: ▷
Typical plumbing system 4
Plumbing tools 72–77

SNAKING A SEWER LINE

Before you rent a sewer cleaning machine, get several bids from professional drain cleaners. Often drain cleaning companies can do the job for only a little more than the cost of machine rental.

If you decide to clean your own sewer line, start by removing the cleanout plug from the main line. If the cleanout plug is made of brass and is threaded into a cast-iron fitting, you may have to chisel it out. Use a sharp chisel and tap it into the brass threads in a counter clockwise direction. Continue to tap the chisel until the plug breaks loose. Because brass plugs are nearly always difficult to remove, you may wish to install a plastic plug when finishing the job. If the cast-iron threads are damaged, you can insert an expandable rubber plug in place of the conventional threaded type.

When using a heavy sewer cleaning machine, always wear leather gloves to avoid injuring your hands. Pay particular attention to the load on the motor. If the motor pulls down too far, stop, pull the cable out and start over.

Most sewer clogs fall into two categories: tree roots and collapsed pipes. Run the cable into the sewer until you feel and hear real resistance. Then pull the cable out and examine the blades. If you find hairlike tree roots, continue snaking until you pull the roots out or push them through. If you find dirt on the end of the pilot, stop. Dirt on the pilot suggests a collapsed line or fitting. You will have to dig the line up to make the repair.

ROOT-CUTTING BLADES

Cleaning a sewer line
Monitor the action of the blades by the sound of the motor.

LEARNING HOW TO SOLDER

Soldered joints are often found in runs of copper water supply pipe. If one should need repair, all you will need is an inexpensive torch, the correct solder and flux, and a willingness to try.

Torches

Most hardware stores offer small, inexpensive torch kits. A typical kit will include a replaceable propane bottle, a regulator valve and a flame tip. The best of these kits offer a turbo tip that spins the flame as it exits the tip. Turbo tips are useful because they wrap flames around a pipe so that you do not have to heat both sides of a fitting. Any of these torch kits will work, however, and you should not have to invest more than $20.

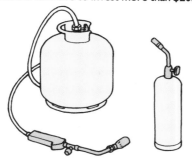

Bottle-type gas torches can be rented but inexpensive, hand-held torches also work well

Flux and solder

The next ingredient you will need is a good, self-cleaning flux. The right flux is critical to achieving a leakproof soldering job. Because you will sometimes need to reflux heated fittings, you will want to avoid soldering paste. Choose, instead, a can or jar of flux that has the consistency of butter. Then buy several acid brushes so that you can brush additional flux onto a hot fitting without burning yourself. Make sure that the flux you buy is self-cleaning. The cleaning agents in self-cleaning flux boil away any chemical residues and tarnish that build up on copper. With self-cleaning flux, you will have to clean only those fittings that are dirty or corroded.

Choosing the right solder is also important. Above-ground residential soldering should be done with 95/5 solder; 50/50 solder has been used for years, but the EPA has recently cited it as a source of potential lead poisoning. Resist any urging by sales people to buy acid-core solder for plumbing.

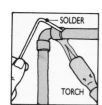

Bending solder
Always bend solder to achieve the greatest maneuverability.

Preparing the joint

Start by making sure that both the end of the pipe and the inside of the fitting are free of dirt and corrosion. If needed, clean the surfaces to be soldered with emery cloth or a wire brush. Then apply flux to the end of the pipe. Flux as much of the pipe as will be covered by the fitting. Then push the pipe into the fitting. When the pipe is fitted and secured with pipe hooks every 4 to 6 feet, you will be ready to solder several joints.

Applying solder

Because comfort and dexterity are important when soldering at odd angles, you will not want to hold solder by the spool. Cut about 2 feet of solder from the spool and wrap all but about 10 inches around your hand. Then pull the loop from your hand and squeeze it into a handle. Finally, bend the last 2 inches of the 10-inch wire at a 90-degree angle. This will give you the greatest maneuverability. The 90-degree angle of the solder wire will allow you to reach all sides of a fitting simply by rotating your wrist. As you use the solder, unwind more from the loop and bend a new angle 2 inches from the end of the wire.

If you are right-handed, place the solder in your right hand and the torch in your left. Light the torch and open the valve all the way. Place the torch tip so that the flame hits the hub of the fitting straight on. The tip should be about ¾ inch away from the fitting. If yours is not a turbo tip, heat one side for a few seconds until the flux begins to liquefy, and then move around the fitting and heat the far side while touching the solder to the fitting.

Keep the solder opposite the torch flame and continue to touch the fitting until the solder liquefies and wraps around the fitting quickly. As soon as the fitting is hot enough to pull solder around it, take the heat away and push solder into the fitting.

If the fitting will not take solder easily, pull the solder away and heat the fitting for a few more seconds. Then push the solder in. A ¾-inch fitting should take about ¾ inch of solder. When the hub has taken enough solder, move onto the next hub on that fitting. Always start with the bottom joint on a fitting and work up. When all joints on that fittings are soldered, watch the rim of the last joint carefully. When the fitting starts to cool, it will draw solder into the joint. This cooling draw is your assurance that the joint is a good one. If the solder around the rim of a joint stays puddled and does not draw in when it cools, heat until the solder liquefies, and then wait again for it to draw in slightly. When you are satisfied that the last joint is a good one, wipe the fitting of excess solder and move on to the next fitting. When all fittings have cooled, turn the water on and check each joint periodically for leaks.

FIRE PRECAUTIONS

Of course, working with a torch requires a few precautionary measures. When soldering a fitting that is next to a floor joist or any combustible surface, you will have to protect that surface. The simplest protection method is to fold a piece of sheet metal over so that it has a double thickness. Then slide this double wall of metal between the fitting and the combustible surface. If you cannot sufficiently protect the area around the fitting, you may wish to solder that section elsewhere and install it already soldered.

To avoid scorching the rubber washers and diaphragms inside valves, always solder them with the handles turned open. This will defuse enough heat to keep the seals from being ruined. Remember, use only as much heat as is needed to draw the solder evenly around the joint. The most common beginner mistake is too much heat, not too little.

Using a heat shield
When soldering near a combustible surface, use a double-thickness of sheet metal as a shield.

Faucet installation has not changed much over the years. While plastic fasteners and flexible supply risers have made the job a little easier, the process remains much the same. These connections still need to be made carefully.

Replacing a faucet

The most troublesome part of replacing a faucet is getting the old faucet off. Start by shutting the faucet's water supply off, either at the meter or under the cabinet. Turn the faucet on to relieve the pressure. Then, loosen the nuts that connect the riser pipes to the supply lines and the faucet. A basin wrench will help you reach the coupling nuts high up under the cabinet. When these nuts are loose, bend the risers slightly so that they can be pulled out of the faucet and supply fittings.

Next, use a basin wrench to undo the nuts that hold the faucet to the sink. If the faucet is old and the nuts corroded, first spray penetrating oil on the threads. If this does not help loosen the nuts, use a small chisel and hammer and gently tap the nuts in a counter-clockwise direction to break them loose. Once broken loose, back the nuts off with a basin wrench.

In a few cases, even these methods will not free the fastening nuts. If this happens to you, your only recourse is to saw through the nut with a hacksaw blade.

Some faucet styles mount from the bottom and are held to the sink or countertop by a locknut under the handle escutcheon. To remove a bottom-mounted faucet, first remove the handles, then the escutcheons. Escutcheons are usually screwed on and can be removed by threading them counterclockwise. Under the escutcheon, you will find the locknuts. Undo these nuts and the faucet should fall out.

Because years of soap and mineral buildup can leave a ridge around the edge of the faucet plate, you may have to clean the faucet area of the sink before installing a new faucet. A 50-50 mixture of white vinegar and warm water used in conjunction with a single-edged razor blade will help you remove this ridge. Just soak the buildup and scrape it away.

Installing a new faucet

After you've cleaned the sink, set the new faucet in place with the rubber or plastic spacer between the faucet and sink. Then reach under the sink and thread the new washers and nuts onto the faucet until the nuts are finger-tight. Before tightening the faucet nuts, however, go back and straighten the faucet so that the back of the coverplate is parallel with the back edge of the sink. When the faucet is straight, tighten the nuts with a basin wrench.

When the faucet is fastened in place, you will be ready to reconnect the supply risers. It is usually a good idea to start with new risers. Some municipal codes allow PB riser pipes. If PB risers are permitted in your locale, use them. Because they are flexible and can be cut with a knife, you will find them much easier to use. PB risers, like copper or chrome risers, include a bulb-shaped head that fits into the ground joint surface of the faucet's hot and cold water inlets. The riser nuts slide onto the risers from the other end. When you tighten the riser nuts to the faucets, the fitted end of the riser is pressed into the ground joints and conforms to make a watertight seal. The supply line ends of the risers are connected with compression fittings.

Basic faucet repairs

Faucet repair is as simple or as complex as the design of the faucet involved. In general terms, there are four basic faucet mechanisms in use today. The oldest type, still used in many faucets and in most valves, uses the stem-and-seat principle. More recent faucet designs feature replaceable cores, rotating balls and ceramic disks. With each of these more recent faucet types, the internal mechanism turns or rotates until holes in the mechanism align with holes in the faucet, allowing water to pass through. The degree of alignment determines the mixture of hot and cold water. In each case, repair is relatively simple and inexpensive. Only when a faucet body itself is defective is faucet replacement absolutely necessary. If your faucet falls apart, replace it.

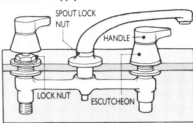

COUPLING NUT
BASIN WRENCH
SUPPLY
SHUTOFF

Disconnect supply with a basin wrench

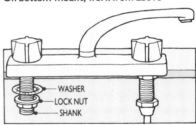

SPOUT LOCK NUT
HANDLE
LOCK NUT
ESCUTCHEON

On bottom-mount, work from above

WASHER
LOCK NUT
SHANK

On top-mount, work from below

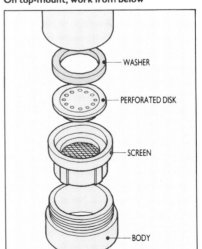

WASHER
PERFORATED DISK
SCREEN
BODY

Uneven pressure
When one faucet has less pressure than others, clean its aerator.

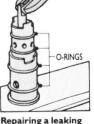

O-RINGS

Repairing a leaking spout collar
Cut the O-rings and replace them.

STEM-AND-SEAT FAUCETS

INDEX CAP

HANDLE SCREW

HANDLE

BONNET NUT

STEM

O-RING PACKING

STEM THREADS

• **Heat-proof grease**
Before reassembling the faucet, cover all moving parts with heat-proof grease. As the name implies, heat-proof grease will not dissolve in hot water. It greatly increases the life of replacement parts and makes the faucet much easier to take apart the next time service is needed. Remember to grease the handle sockets as well. Heat-proof grease will separate dissimilar metals enough to prevent corrosion.

SEAT WASHER

WASHER SCREW

BODY

Typical stem-and-seat faucet

Replacing stem packing

PACKING STRING

Replacing stem packing
Wrap packing string around the stem and tighten the packing nut.

If a faucet leaks around the handle when you turn the water on, the packing washer is defective. Before replacing the stem, packing gland, packing nut and handle, check the packing washer. Packing washers are made of graphite, leather, rubber or nylon, depending on the age and make of the faucet. When a packing nut or packing gland is tightened, it compresses the packing washer which seals against the stem. If your faucet has a visible brand name, you can buy replacement packing washers. If you are unable to find a replacement washer, or if your faucet's original packing was made of packing string, then your best alternative is to repack the stem with graphite or plastic packing string. Simply wrap the stem below the nut several times and thread the nut in place. The string will compress and fill in any voids against the stem.

Repairing stem-and-seat faucets

To repair stem-and-seat faucets, start by removing the handles. Most handles have decorative coverplates under which you will find handle screws. These coverplates are also called "index caps" because they indicate hot and cold water. Pry the caps off and remove the screws. If a handle has not been removed recently, it may be stuck to the stem. Gently pry up under both sides of the handle with two screwdrivers to free the handle. If the handle will still not loosen, you may need to buy an inexpensive handle puller from your local hardware dealer.

On better faucets, all external parts are made of chromium-plated brass. Brass faucets usually come apart easily. Manufacturers of more competitive faucets, however, often substitute chrome-plated pot metal for brass on handles, coverplates and escutcheons. Pot metal will corrode, through electrolysis, when in contact with dissimilar metals. Pot metal handles, in particular, are often difficult to remove and are easily damaged when force is applied. If you must damage a handle to get it off, your best bet is to replace it with a universal-fit replacement handle.

When you have the handle off, undo the escutcheon so that you can get to the packing nut and locknut. When you

reach the locknut or packing gland, loosen it with a thin adjustable wrench. If the locknut backs out several rounds and then stops, turn the stem in or out, depending on the brand, to free the locknut. When the locknut is loose, back the stem out of the faucet. Most stems will turn out counterclockwise.

On the end of the stem you will find a rubber washer secured by a brass screw. Undo this screw and find a replacement washer that fits the rim of the stem. A tight fit is important here. The washer should not be too big or too small. Press the washer in place and replace the screw. Because brass screws can become brittle with age, you may wish to replace the washer screw as well. Washer screws often come with washer assortment packets.

The kind of seat your faucet has will determine the kind of replacement washer you will need. Recessed, beveled seats require beveled washers. Seats that have raised rims require flat washers to make a seal. When in doubt as to which type of washer you should use, check the shape of the seat. Because someone before you may have repaired your faucet with the wrong shape washer, using an old washer to determine the shape of the new is not a good idea.

DEALING WITH SEAT DAMAGE

When a defective washer is allowed to leak for an extended period of time, the pressure of the water will cut a channel in the faucet seat. For this reason, always repair a leaking faucet immediately. A defective seat will chew up new washers in short order. Always check the faucet seat when you replace a washer.

If you find a channel in the surface of a seat, replace the seat. A seat wrench will allow you to unscrew a removable seat from a faucet body.

Some seats are machined into the brass body of a faucet and therefore cannot be removed. If your faucet seats are pitted and cannot be replaced, your only alternative is to grind the entire seat rim to a level below the surface of the pit. While this may sound difficult, it is not. You can buy an inexpensive seat grinder from your local hardware dealer and once the faucet is apart, you can complete the job in a matter of minutes.

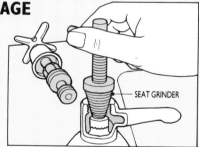

SEAT GRINDER

Damaged seats can be refurbished

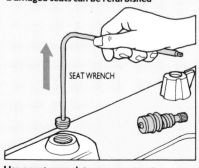

SEAT WRENCH

Use a seat wrench to remove a seat

SEAT-AND-SPRING FAUCETS

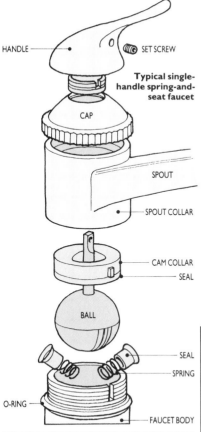

HANDLE — SET SCREW

Typical single-handle spring-and-seat faucet

CAP

SPOUT

SPOUT COLLAR

CAM COLLAR
SEAL

BALL

SEAL
SPRING

O-RING

FAUCET BODY

Repairing handle leaks

If your seat-and-spring faucet leaks around the handle but does not drip, all you have to do is tighten the body cap. The cap is the threaded dome that holds the mechanism in place. The repair kit comes with a small cap wrench, or if you prefer, you can use a slip-joint pliers to tighten a cap. You simply slip the wrench onto the cap and turn it clockwise until the leak stops. Do not overtighten. If you use a pliers on the knurled surface of the cap, wrap the jaws of the pliers with a soft cloth to avoid stripping chrome from the cap.

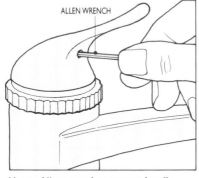

ALLEN WRENCH

Use an Allen wrench to remove handle

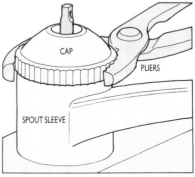

CAP
PLIERS

SPOUT SLEEVE

Use pliers to turn the cam cover

Seat-and-spring faucet repairs

Seats-and-spring type faucets, one of a variety that have come to be known as 'washerless' faucets, have become quite popular in recent years, both because of their durability and because they are easy to repair. The operating mechanism of a seat-and-spring faucet consists of a stainless-steel or plastic ball that is turned and aligned with water openings in the faucet body as the single handle is manipulated. The openings in the faucet body contain spring-loaded rubber caps, or seats, which press against the globe and prevent leaking when in the off position. When this type of faucet drips, it is because these rubber cups and springs are worn.

Repair kits for seat-and-spring faucets are inexpensive and come with everything needed to completely rebuild a faucet. It is important to identify the faucet by brand name when asking for a repair kit in order to get the correct parts. The common brand names are Delta and Peerless. Kits come with complete instructions. The repairs you make will depend upon the location and nature of the leak.

REPLACING SEATS AND SPRINGS

If the faucet drips, you must take the faucet apart. This is easily enough done by using the proprietary cap wrench. Use the Allen wrench side of the tool to loosen the handle set screw. Pull the handle off. Then use the cap wrench to loosen and remove the cap. Under the cap you will find a nylon and rubber cam covering a stainless-steel ball. Remove the cam assembly and pull the ball up and out of the faucet body. Inside the faucet you will see two hollow rubber caps, or seats, mounted on two small springs.

Insert a needlenose pliers into the faucet body and pull the seats and springs out. Throw the worn seats and springs away. It is too easy to get the old confused with the new. Then slide the new springs and seats onto a screwdriver and slide them into the faucet holes. When the spring-loaded seats are in place, you will be ready to replace the ball-and-cam assembly.

The kit will come with a new cam seal as well. The cam assembly will have a tab on one side that corresponds with a slot in the faucet. Match the tab with the slot and press the cam assembly in place over the ball. You can then tighten the cap and replace the handle.

If your seat-and-spring faucet also leaks around the spout collar, you should replace the collar O-rings before putting the cap and handle back on. To replace collar O-rings, pull up evenly and firmly on the spout until it comes free. Use a knife to cut the old O-rings from the slots and slide the new rings over the body until they fit into the O-ring slots. Then grease the rings with liquid detergent or heatproof grease and press the spout collar back on, rotating it gently as you go.

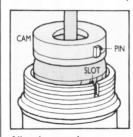

CAM
PIN
SLOT

Align the cam pin

SEAL

Retrieve seats and springs

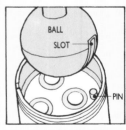

BALL
SLOT
PIN

Align the slot in the ball

WASHERLESS FAUCETS

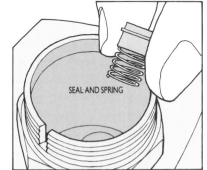

LOCK NUT

Remove the locknut to reach the cartridge

INDEX CAP

SEAL AND SPRING

Pry under cap

Replace defective seats and springs

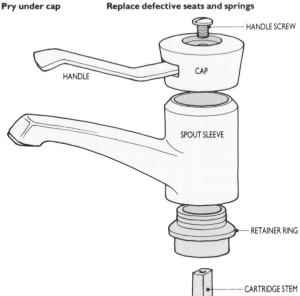

HANDLE SCREW

HANDLE

CAP

SPOUT SLEEVE

RETAINER RING

RETAINER CLIP

CARTRIDGE STEM

O-RING

FAUCET BODY

O-RING

Typical single-handle cartridge-type faucet

Repairing washerless two-handle faucets

Washerless faucets are also available in two-handle designs. These faucets are also easy to repair. The repair kit includes a cylinder, or cartridge, and seat assembly, which is similar to the seat-and-spring combination used in the single-handle washerless faucets. These cartridges come with new O-rings already installed.

To repair a two-handle washerless faucet, pry the index cap from the handle and remove the handle screw. Plastic two-handle washerless handles come off easily. When the handle is off, use a wrench to undo the locknut that holds the cartridge in place. Then pull the old cartridge assembly straight up and out of the faucet. Replace the seat and spring as you would with a single-handle faucet.

The new cartridges must first be properly aligned before they can be installed. On the side of each cartridge you will see a tab and a slot separated by an O-ring. The tab is called a key and its corresponding slot is called a keyway. You will also see a raised stop at the top of the cartridge next to the handle stem. Align the key so that it is directly over the keyway. Then insert the cartridge so the stop is facing the spout. If you are replacing hot and cold cartridges, both stops should be facing the spout. When the cartridge is properly positioned, press it into the faucet. Replace and tighten the locknut until it is snug. Press the handle onto the new stem and replace the handle screw and index cap.

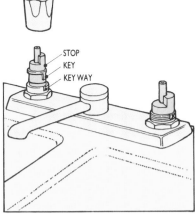

STOP
KEY
KEY WAY

Align keys and keyways

Repairing cartridge-type single-handle faucets

Cartridge-type washerless faucets have also enjoyed wide popular use because of their simplicity and durability. They are available in single-handle and two-handle versions. Unlike the single-handle seat-and-spring mechanism, the cartridge type has a core instead of a ball. This type of faucet is marketed widely under the Moen trade name.

All single-handle Moen faucets use the same core mechanism, but other aspects vary with price. The major differences are in exterior construction and style. In terms of repair, the only notable difference is in the way the spout/handle covers come off.

If your Moen single-handle faucet has a flat chrome coverplate with the trade name pressed into it, you must first pry the coverplate up with a knife to get to the handle screw. Other models have a plastic hood that covers the handle screw and top of the faucet. These hoods simply pull up and off. In the case of lavatory faucets, the handle screw is reached by prying off a plastic index cap at the top of the handle.

Below the handle you will find the core stem. If yours is a chrome coverplate model, you will also need to remove a retainer pivot nut. Just below the stem, or pivot nut, you will see a brass clip inserted into the side of the cartridge stem. This clip locks the cartridge in place. Assuming you have the water turned off, pull the clip out of the faucet with pliers. When the clip is removed, you will be able to pull the cartridge out of the faucet body.

Press the new core in place, making sure that the keys of the cartridge are aligned front to back. Seat the cartridge by pressing on the keys until the cartridge is in far enough to accept the clip. Then slide the clip back into the faucet body until it grips the far side of the faucet wall. Then replace the retainer pivot nut and handle. Make sure that the handle slips into the groove in the retainer pivot nut before replacing the screw. You will know that the handle is working properly when it lifts and lowers the stem of the cartridge smoothly. If the spout has been leaking when the faucet is in the on position, you should replace the spout collar O-rings before replacing the handle.

Repairing two-handle Moen faucets

Two-handle Moen kitchen and bath faucets also use cartridges for their internal mechanisms. The cartridges for two-handle faucets are smaller than single-handle cartridges of the same make. They are easy to install.

To remove a defective two-handle cartridge, start by removing the handle. The index cap will pry off with a thin-bladed knife so that you can get at the handle screw. Under the handle you will find a large nut that holds the cartridge in place. Undo this nut with pliers (1). Then lift the cartridge out by

the stem, pulling straight up.

To install a new cartridge, turn the cartridge stem in a counterclockwise direction (2) so that the holes of the cartridge are aligned (3). Then push the cartridge straight down into the faucet, making sure that the key at the top of the cartridge fits into the slot in the faucet. When you've seated the cartridge, screw the cartridge nut back on until it is hand-tight. Tighten with pliers until snug. Be careful not to strip the threads. Then replace the handle, handle screw and index cap.

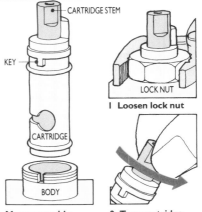

Moen assembly

1 Loosen lock nut

2 Turn cartridge stem

Repairing ceramic-disk faucets

When a ceramic-disk faucet leaks, it is likely to show up around the base of the faucet or on the floor below. Getting to the operational part of these faucets is a little different than with other models. Instead of prying off the index cap to remove the handle, you must tip the handle back to reveal a set screw under the front of the handle. Use an Allen wrench to loosen the screw. But having the handle removed still does not give you access to the internal mechanism. You must also remove the chrome faucet cover. With older models, you will need to loosen the pop-up drain lever and undo two brass screws from the underside of the faucet. Newer models have a slot screw in the handle

and a brass keeper ring that allows you to remove the cover from above.

When the faucet cover is off, you will find a ceramic disk secured by two brass bolts to the base of the faucet. Undo these bolts and the disk should lift off. Take this disk with you to your plumbing outlet and buy an identical replacement disk.

To install the new disk, align the ports of the disk with those of the faucet base and make sure that the flange under one of the cartridge bolt holes fits into the rim around one of the bolt holes in the body plate. When all is perfectly aligned, replace the disk bolts and refasten the faucet cover and handle. Then turn the water on and test for leaks.

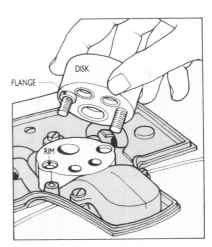

3 Remove body to reach the disk

Sillcock repair

A sillcock is simply a stem-and-seat faucet with a long stem. Depending upon the length of the drain chamber, the stem can be from 6 to 30 inches long.

Start by shutting off the water at the meter. With the stem in the open position, loosen the locknut and pull the stem straight out. If the stem is particularly stubborn, twist and pull it out with a locking pliers. When you have the stem out, replace the washer, apply heat-proof grease and replace the stem.

Sillcocks often wear out faster than ordinary faucets because of a common error in judgment. As water continues to drain from the chamber after the flow has stopped, many homeowners continue to turn the handle. This extra, unnecessary pressure ruins stem washers in a hurry.

REPLACING FREEZELESS SILLCOCKS

Freezeless sillcocks differ from ordinary outside faucets in that they can be left on during freezing weather. Instead of stopping the water on the outside of the home, they stop water inside the home by means of a long faucet stem. The only way a freezeless sillcock can freeze is if a hose is left connected in cold weather or if it has been installed without sufficient pitch to drain. When either situation occurs, sillcocks will freeze and split just inside the home near the valve seat.

If a damaged sillcock is not buried in a basement ceiling, replacing it will be easy. First, shut the water off and undo the screws from the outside flange. Then, unthread the sillcock from the pipe in the basement. If it is soldered to a copper line, you will have to cut the line to pull the sillcock out. With the old sillcock out, buy a new one the same length and thread it onto the supply line. Copper connections, of course, will require soldering a splicing coupling between the cut pipe and sillcock.

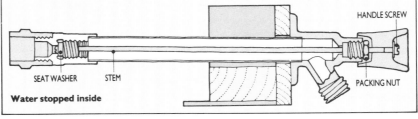

Water stopped inside

● **Replacing a concealed sillcock** Often sillcocks are buried in walls and ceilings. To avoid a troublesome wall or ceiling repair, measure carefully to determine the exact location of the connection and cut a 6-by-10-inch hole in the drywall or plaster. Make the change through this hole and then cover it with a furnace grill. The grill will successfully hide the opening and the louvers will allow warm air in around the sillcock, thus preventing a future freeze. Because a sillcock must drain any water left between its seat and hose connection, make sure that you install yours with enough pitch to drain the chamber completely.

19

LAVATORIES AND SINKS

Replacing a wall-hung lavatory with a vanity cabinet and countertop basin will greatly improve the appearance of your bathroom while adding valuable storage space. This is one project that will add real value to your home with less than a weekend's time invested. The amount you spend on materials will depend on the size and construction of the cabinet and the type of faucet and basin you choose.

How to replace a wall-hung lavatory with a vanity and basin

The first step is to measure the space around the existing lavatory to determine what size cabinet your bathroom will allow. Factory-made cabinets and countertops come in standard dimensions to fit many opening widths and depths. Buying one of these cabinets is generally the less expensive alternative to having one custom made. If the space you have will not accept a standard size cabinet, your next option is to have a cabinet built to fit the available space.

Countertops and lavatory basins are other major considerations. For standard-size cabinets, cultured-marble tops with basins molded into them are a good choice. Another option is to mount a china, cast-iron, plastic or enameled-steel basin in a particleboard deck that has been finished with a decorative plastic laminate. Countertop coverings are available in a variety of textures and colors to match nearly any home style or decor.

If your cabinet is custom made for a nonstandard dimension, a cultured-marble top is not likely to fit. Beyond this restriction, the combination you choose will be a matter of taste and adhering to your budget.

Removing a wall-hung lavatory

To remove a wall-hung lav, start by disconnecting the trap and water risers. If your lav does not have shutoff valves between its risers and supply lines, now might be a good time to install them. With stops on both supply lines, you will not have to shut the entire system down while you work.

When the piping has been disconnected, check to see if the back of the lavatory is secured to the wall by anchor screws. These screws will be located at the back of the lav just under the apron. If anchor screws are present, remove them. Then grab the lav by its sides and pull up. It should lift right off the mounting bracket. Remove the mounting bracket by undoing the screws that fasten it to the wall.

Before installing the new vanity, check to see if the wall above the vanity needs to be repaired or painted. Often the screw holes from the mounting bracket will show above the new vanity. Filling these holes and other cosmetic repairs may be necessary before the vanity can be set in place.

With the opening ready, slide the new vanity in place and secure it to the wall with long screws through the back of the cabinet. In some cases you may have to cut openings for the drainpipe and supply lines. With the cabinet in place, you are ready to set the countertop or premolded top and basin. In the case of a cultured-marble top, simply set it on the vanity and glue it to the cabinet top with construction adhesive. In the case of a factory or custom-made countertop, screw the top to the corner brackets of the cabinet from the underside.

Cutting a basin opening

The lav basin you buy will come with a paper template. Position this template where you want it on the countertop and tape it down. Then use a jigsaw to cut the opening for the basin along the dotted line of the template. To avoid chipping the plastic laminate, use a fine-tooth blade and advance the saw with steady, even pressure.

Because it is much easier to install a faucet and pop-up assembly sitting down than while on your back in a cramped space, you will want to fasten the faucet and pop-up to the basin before installing the basin in its countertop opening.

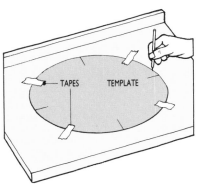

Trace around the paper template

INSTALLING A FAUCET AND POP-UP

To install a lavatory faucet, insert the faucet supply shanks through the basin faucet holes. Then slide the large spacing washer onto the shanks from below and tighten the locknut on each shank. Make sure that the faucet is centered before tightening the locknuts.

To install a pop-up drain assembly, pack putty around the drain flange and thread the pop-up waste pipe into the basin gasket until the flange is seated in the drain opening. Follow by threading the tailpiece into the pop-up waste pipe.

To connect the pop-up mechanism, insert the lift rod into the opening at the back of the faucet. Then slide one end of the adjustment lever onto the lift rod and the other onto the pop-up lever. Finally, pull the pop-up lever all the way down and tighten the adjustment screw.

When the basin is in place and hooked up to its trap and water supplies, use water-soluble latex caulk to seal the basin to the counter. Wet the areas to be caulked first. Then apply a bead of caulk around the faucet base. Push caulk into the cracks. Then wipe all excess caulk away with a damp cloth.

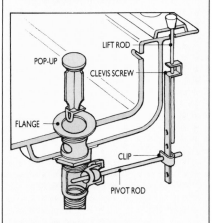

Typical pop-up drain assembly

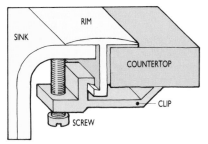

Typical sink rim fastening details

Removing a wall-hung lavatory
Removing a wall-hung lavatory
Pull the lavatory off the wall bracket.

REPLACING A KITCHEN SINK

Replacing a kitchen sink is not a difficult task. How you proceed, however, depends upon whether or not you replace the countertop as well.

Removing the old sink

If you plan to replace your old countertop, there is probably no need to pull the sink from the counter. Simply disconnect the trap, disposer and water supplies, remove the screws from the underside of the countertop and pull the sink and countertop up together. If you wish to save the basket strainers, disposer flange or faucet, these are much easier to remove when the countertop has been removed.

If you plan to save the countertop and replace only the sink, first remove the old sink without damaging the countertop. Start by disconnecting the water supplies, trap and disposal. Then loosen the clips around the underside of the sink rim. A special tool, called a Hudec wrench, will allow you to reach these clips with less strain, but a simple nut-driver socket will work.

When the clips have all been removed, lift the sink straight up by placing one hand around the faucet base and another through the disposer opening. With the sink out, you will likely need to clean a mineral, putty and soap buildup from the counter where the old sink rim rested. A putty knife and household cleanser work well in removing any buildup, but work carefully to avoid scratching the surface.

Replacement sinks

The replacement sink you choose will have one of three possible rims. If you select an enameled cast-iron sink, you will be able to choose between a "self-rimming" model and one that requires a sink rim to fasten it to the countertop. The "self-rimming" type has a rolled edge that rests on top of the counter and is caulked in place. The edge of the sink is raised and sets above the counter. The disadvantage of a self-rimming cast-iron sink is that when wiping the counter, you cannot simply push spills and food particles into the sink. As a result, some homeowners prefer a cast-iron sink that fits flush with the countertop. For a flush fit, choose a sink that uses a sink rim to suspend it in the counter opening. Sink rims are also required on porcelain steel sinks. A third rim seal is found on stainless-steel sinks. No separate rim is needed but rim clamps are used to fasten the sink to the countertop.

Cutting into a new countertop

Most new sinks will come with a paper template to help you make the right size cut in a new countertop. Position this template on the counter over the sink cabinet so that it is centered. Try to leave at least ½ inch between the sink rim and the backsplash. Then tape the template down and make the cut through the paper with a saber saw. It helps if you can have someone hold the section being cut from below so that it does not fall through and crack the countertop.

If you wish to install a used sink in a new countertop, you will have to do without a template. You can measure the dimensions to be cut, but it is easier and more exact to lay the sink rim on the counter and trace around it. Remember that your cut will need to be made from the center of the rim so that the outside of the rim can grip the countertop. If you are installing a self-rimming sink without a template, remove the faucet and lay the sink upside down on the counter. Trace around it lightly with a pencil. Remove the sink and draw a new line ½ inch inside the traced line. Then make the cut on the inside line.

SEE ALSO

Details for: ▷	
Installing faucets	15
Plumbing tools	72–77

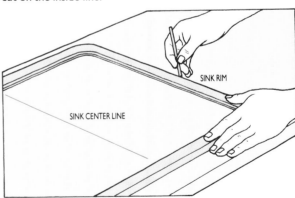

Set rim in place on counter and trace around it

Mounting a stainless-steel sink

Stainless-steel sinks are available in a wide range of prices. The more shiny the surface, the more expensive the sink. Stainless-steel sinks do not require sink rims to hold them in place. Instead, the rim of the sink rests on the top of the counter and is held against the countertop by fastening clips from the underside of the counter. Stainless-steel sinks will usually fit the same opening as a sink rim and therefore make good replacements. Measure before you buy.

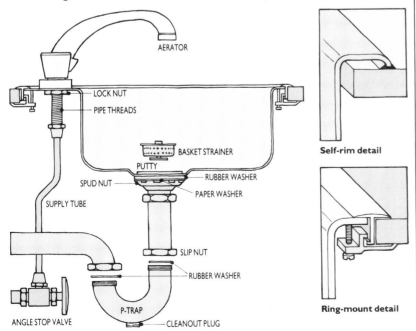

Typical sink installation details

Self-rim detail

Ring-mount detail

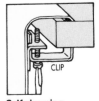

Self-rimming stainless-steel sink mounting detail

21

STRAINERS AND DISPOSERS

The chrome trim you see in the drains of your sink are not part of the sink basin. They are merely the most visible parts of your disposer and drain spud. As such, they can be replaced when cosmetic or mechanical problems arise. The job is involved but not difficult and requires only a few tools.

Replacing a basket strainer/drain spud

To remove a basket strainer, disconnect the drainpipe from the basket drain spud. Then use a spud wrench or a large adjustable pliers to loosen the locknut. To keep the strainer body from turning when you loosen the locknut, insert the handles of a small pliers into the drain crosspiece from the top. While you turn the nut from below, have someone hold the pliers with a wrench or screwdriver from above.

If the nut is corroded on the strainer body, you may need to use a small chisel and a hammer to break the threads loose. If the hammer and chisel do not loosen the threads, use a hacksaw blade to cut the locknut. Tape one end of a hacksaw blade to create a handle. Cut across the nut in an upward, diagonal motion.

When the locknut is loose, push the strainer body up through the drain opening. Clean the brittle, old putty from the recessed flange of the sink opening. Then roll new putty out between the palms of your hands so that the putty forms a soft rope about ½ inch in diameter. Press this putty around the flange of your new strainer and press the strainer in place.

Then slide the rubber washer onto the spud from below. Next, slide the fiber washer on, followed by the new locknut. Tighten the locknut until it no longer turns or until the strainer body also begins to turn. From above, clear the excess putty from the rim of the strainer and tighten the locknut again.

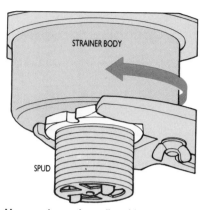

Use a spud wrench or adjustable pliers

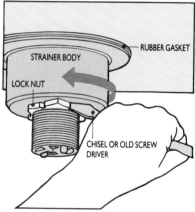

If the spud nut sticks, tap it loose

Installing a disposer in an existing sink

A disposer can be installed in any sink that has a full-size drain opening. All you will need is a disposer waste kit and access to electricity.

To install a disposer in an existing two-compartment sink, you must first remove the waste connector that joins the two drains to the P-trap. Then remove the strainer and clean any remaining putty from the recessed sink flange.

With the basket strainer removed, you are ready to install the disposer sink flange. Press putty around the sink flange and press the flange into the recessed sink opening. Then slide the rubber gasket, fiber gasket and mounting ring onto the spud from below. While holding the flange in place, snap the retaining ring over the ridge on the mounting ring. Tighten the screws on the mounting ring until nearly all putty is forced from between the sink and flange, then trim the putty.

With the flange mounted in the sink opening, attach the disposer. Some disposers fasten to the flange by means of a stainless-steel band. Hold the disposer up so that the rubber collar on the disposer fits over the flange. Then tighten the stainless-steel band with a nut-driver or screwdriver.

Other models attach to their flanges by means of fastening rings. If the model you buy has this type of fastener, hold the disposer up to the flange and turn the fastening ring until its guides catch the tabs on the mounting ring. Then turn the fastening ring until it seats over the ridges on the upper mounting ring.

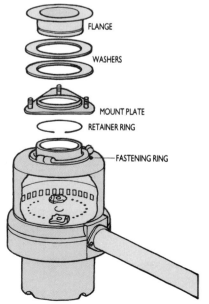

Typical disposer flange installation

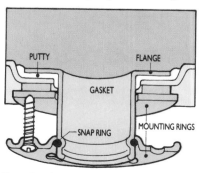

Fastening ring assembly

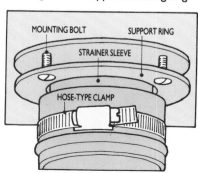

Hose clamp connection

CONNECTING TO WASTE LINES

Most disposers on the market today come with a discharge tube that connects directly to the P-trap under your sink. These discharge tubes, however, work only if you are installing a disposer in a single-compartment sink or if you are replacing a disposer that was previously hooked up to its own trap.

When connecting a disposer under a two-compartment sink, both compartments can be drained through a single trap. To make this connection, do not use the discharge tube that came with the disposer. Instead, buy a disposer waste kit. This kit consists of a baffled tee, a rubber gasket and a flanged tailpiece. Install the tee vertically between the P-trap and the tailpiece extension from the other compartment. The center of the tee branch opening should be only slightly lower than the discharge opening of the disposer.

With the tee installed at the proper height, slip the rubber gasket onto the flange of the tailpiece and hold it between the disposer opening and the tee opening. The tailpiece will be longer than needed, so you will have to cut to fit. Insert the flange in the disposer discharge opening and hold the other end of the tailpiece against the hub of the tee to determine the desired length. Remember to include the depth of the hub in your measurement.

Next, slide the metal flange onto the tailpiece followed by the compression nut and compression ring. Insert the compression end in the hub of the tee and the gasket end into the disposer. Bolt the flange to the disposer and tighten the compression nut to the tee.

To keep cooking grease from clinging to the sidewalls of the disposer and drain, always run cold water through your disposer when the motor is on. Cold water causes grease to coagulate and flow through the pipes. Hot water thins grease and allows it to build up in pipes.

WARNING

Never make electrical connections until the power is switched off at the service panel.

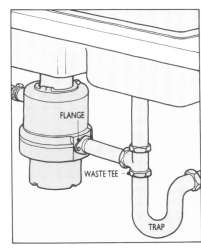

Typical disposer installation

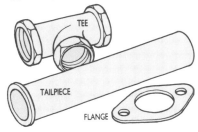

Components of waste kit

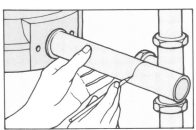

Mark and cut tailpiece

Waste flange tailpiece

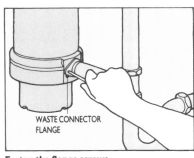

Fasten the flange screws

ELECTRIC HOOKUPS

If you are installing a disposer in a sink that has not had a disposer before, you will have to find some way to get electricity to the disposer and to a switch near the sink. If the basement ceiling beneath your sink is not finished, the easiest alternative might be to run a separate cable from your service panel into the wall behind your cabinet to a switch and then back down into the cabinet to your disposer. To meet code requirements, however, the leader wire in the cabinet must be encased in flexible conduit.

If this option is not available to you, then you may wish to convert an existing outlet box above your counter into a switch box and fish a short length of wire down the wall and into the cabinet to the disposer.

Once inside the cabinet, the electrical hookup is simple. Remove the coverplate from the bottom of the disposer and pull out the white and black wires. Then, using plastic insulating connectors, connect the two insulated wires from the switch box to the insulated wires in the disposer. Connect black to black and white to white. Then fasten the uninsulated ground wire from the switch to the ground screw inside the disposer. Replace the coverplate and test the disposer.

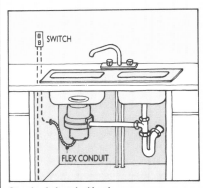

Standard electrical hook-up

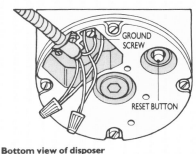

Bottom view of disposer

SEE ALSO

Details for: ▷	
Plastic drain fittings	9
Joining dissimilar materials	10

● **Servicing your disposer**
Disposers work very well on most foods but have real problems with anything hard or stringy. Chicken bones, fruit seeds, eggshells and celery can stop a disposer cold. When the motor pulls against too much resistance, a safety breaker built into the disposer will trip, cutting the power to the motor. To get a stopped disposer started again, first dislodge the motor blades from the blockage. To do this, use the wrenchette that came with the disposer to reverse the motor manually. The wrench should fit into a key slot on the bottom side of the disposer. Turn the motor back and forth until it spins easily in both directions. Next, find the reset button on the underside of the disposer and press it. Finally, run cold water through the disposer drain and turn the disposer on. This wrench and reset procedure will free most blockages. If the disposer stops again, repeat the procedure.

INSTALLING A BUILT-IN DISHWASHER

Like all things mechanical, dishwashers wear out. Hiring someone to replace your old dishwasher can easily add $100 to the price of a new one. Dishwashers are not terribly difficult to install, but one *precaution is in order. Because you must work with both electricity and water, make sure that you shut the electricity off at the main service panel before beginning dishwasher removal or installation.*

SEE ALSO

◁ Details for:
**Draining a
water system**   6

Removing your old dishwasher

After shutting off the electrical current to your old dishwasher, look under the sink for the valve that shuts the water off (◁). In many cases this valve will be an in-line globe valve that is soldered or threaded in place. If you find no valve isolating your dishwasher, you will have to shut off the supply to the entire house at the meter.

With the water shut off, disconnect the discharge hose from the garbage disposer or the dishwasher tailpiece under the sink. In each case, the discharge hose will be held in place with a hose clamp. Simply undo the clamp and pull the hose out.

Next, remove the cover panel beneath the door of the dishwasher. This will give you access to the water and electrical connections. You may have to open the door to gain access to these screws. When the screws are removed, the panel should lift off.

Under the dishwasher, on the left-hand side near the front, you will find the water supply pipe and fitting attached to the solenoid valve. In most cases, the supply pipe will be ⅜-inch soft copper and the fitting will be a ½-inch threaded male iron pipe by ⅜-inch compression angle adapter. This fitting is used so universally on dishwashers that it is called a

"dishwasher ell." Loosen the compression nut and pull the copper tubing out. Keep a shallow cake pan handy to catch the water trapped in the line. Leave the copper tubing in place under the dishwasher, in most cases it can be connected directly to the new dishwasher unit.

Near the solenoid valve you will see the metal box that contains the electrical connection. Undo the screws from the coverplate and pull the wires out of the box. Undo the connections of the three wires and save the plastic connectors for the new installation.

Most dishwashers are fitted with brackets that fasten to the underside of the countertop. Remove the screws from these brackets, but before you pull the machine out, turn each of the four leveling legs in so that the top of the dishwasher will clear the edge of the countertop. Then tilt the dishwasher back and slide a large piece of cardboard under both front legs. This will keep you from tearing or gouging your floor covering. With everything disconnected and the cardboard in place, ease the old dishwasher out. Before hauling the old dishwasher away, however, tip it up, and remove the dishwasher ell from the water supply inlet. You may need it for your new supply connection.

Installing your new dishwasher

The dishwasher you buy will come in a cardboard box and will be bolted to a wooden skid. Cut the box around the bottom and lift it off. Then tilt the dishwasher up and undo the lag bolts from the skid. Remove the access panel and thread the dishwasher ell into place. Be sure to use pipe joint sealant tape or pipe joint compound on the male threads of the ell. Tighten the ell until nearly snug and then tighten one more round until the compression nut points in the direction of the existing water supply pipe.

Slide the dishwasher up to the cabinet and start the drain hose through the existing opening at the back of the cabinet. Slowly push the dishwasher into place. As the dishwasher goes in, have someone pull the discharge hose into the sink cabinet.

Align the sides of the dishwasher evenly in the cabinet. Make sure that the front of the dishwasher (not the front of the door) is flush with the cabinet. Then extend the leveling legs so that the fastening brackets meet the bottom of the countertop. Open and shut the door several times to test it. If the door rubs against the cabinet stiles, adjust the position of the dishwasher and then fasten the brackets to the countertop with short screws.

TESTING FOR LEAKS

Before replacing the access panel, run the dishwasher through an entire cycle so that you can check your work. If the frame of the dishwasher vibrates too much, you may need to adjust one of the leveling legs. If a small amount of water appears beneath the water connections, a quick tightening of compression nuts will usually correct the problem. Be careful not to overtighten. If the pump motor or electrical features fail to work, check the fuse or breaker.

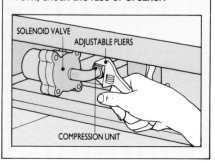

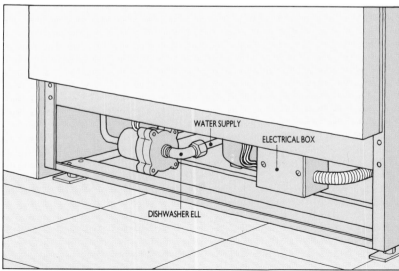

Typical dishwasher installation

WATER SUPPLY

ELECTRICAL BOX

DISHWASHER ELL

SOLENOID VALVE

ADJUSTABLE PLIERS

COMPRESSION UNIT

CONNECTING A DISHWASHER

Connecting the discharge hose

A discharge hose has an inside diameter of ⅝ inch, which is smaller than the fitting on a garbage disposer. You will need to compensate for this difference by installing a dishwasher adapter kit between disposer and discharge hose. This adapter is merely a rubber fitting that enlarges in ½-inch increments from ½ inch to 1⅛ inches. Each end is banded with hose clamps. Cut the adapter at the approximate size and insert the hose directly into it. Then band the adapter to the hose.

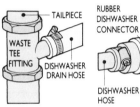

In-line dishwasher tailpiece

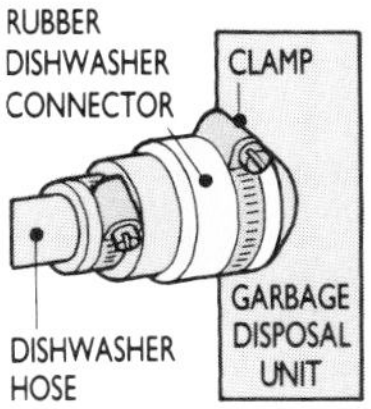

Dishwasher waste kit

The most important thing to remember when hooking up a discharge hose, however, is that the hose must arch high up in the cabinet before descending to the disposer or dishwasher tailpiece. If your kitchen sink ever backs up, bacterial-laden sewage will flow into your dishwasher if no loop is there to stop it.

Some codes may require that a vacuum breaker be installed in the discharge hose. A vacuum breaker provides an air gap that prevents siphoning, and because the hose connects to it just under the countertop, it creates its own loop. The most likely spot to install a vacuum breaker is in the fourth hole in your sink. If no extra sink hole exists, cut a hole in your countertop with a hole saw. The top half of the vacuum breaker will fit on the top and the hose attachments will fit under the counter.

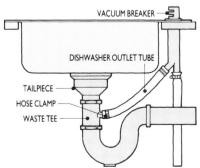

Some codes require a vacuum breaker

Making the water connection

If the previous water connection was made in ⅜-inch O.D. or I.D. soft copper pipe, the new connection will be easy. With the dishwasher ell transferred from the old dishwasher to the new, all you will have to do is apply pipe joint compound to the ferrule already on the pipe, insert the end of the copper into the ell and tighten the compression nut.

If your old dishwasher was piped in rigid copper with soldered joints, the process becomes a little more complicated. Rigid copper supply pipes enter the dishwasher compartment in one of two ways. The most common approach is directly through the floor. In this case a union is located under the dishwasher, and a valve is located just below the floor in the basement. If the water supply does not come through the floor, then it will take off from the hot water supply line in the sink cabinet, enter the dishwasher compartment and travel along the floor to the dishwasher water inlet. In either case, it is usually best to cut the supply line just after the valve and install a ⅝ × ⅜-inch compression adapter. You will then be able to run ⅜-inch soft copper pipe through the cabinet wall or kitchen floor to the compression end of the dishwasher ell. If manufacturer specifications insist on larger supply lines, use larger soft copper pipe.

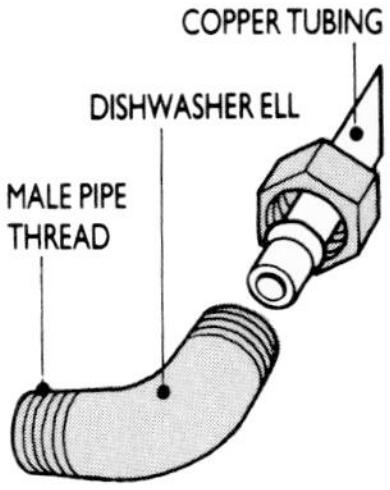

Making the electrical connection

Most codes require that the non-metallic sheathed cable that extends from the wall to the dishwasher within the dishwasher compartment be encased in flexible conduit (▷). If your wire is not, now is a good time to do it. Just slide a length of conduit over the wire so that it enters the drywall. Then cut the other end so that it can be fastened to the electrical box on the dishwasher.

With the conduit in place, use plastic wire connectors to connect the respective wires inside the box. Attach black to black, white to white and fasten the ground wire to the ground screw inside the box.

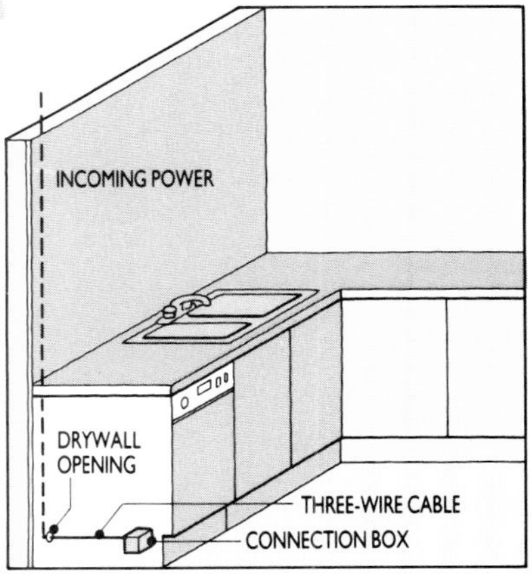

Run a new cable from service panel

COPPER TUBING
DISHWASHER ELL
MALE PIPE THREAD

A dishwasher ell adapts soft copper to a solenoid inlet

KEEP YOUR DISHWASHER WORKING SMOOTHLY

If you are installing your home's first dishwasher, you should also have your kitchen drain line cleaned. Often a partially plugged drain line will accommodate the relatively low output of a kitchen sink, but when a dishwasher is added, the line will overflow. If you snake the line first, you will prevent the chance of water damage. Because a dishwasher forces a lot of water through a drain line and because that water is always very hot, a snaked kitchen line will remain clean almost indefinitely.

Dishwashers are designed to retain some water. Their pumps contain rubber O-rings and seals that must stay wet. Without this water, your dishwasher's seals will dry out, causing the pump to seize up or leak. If you are going to be away from your home for two weeks or longer, pour a thin layer of mineral oil into the base of your dishwasher. The oil will float on top of the water and seal it so that evaporation does not occur.

Contrary to popular belief, dishwashers do not sterilize dishes. The water temperature would have to be at least 180 degrees F to sterilize, and water at that temperature is a safety hazard at your faucets and in your shower. You should set your water heater between 135 and 140 degrees F. To see what your current hot water temperature reads, use a meat thermometer under one of your faucets. Make sure you hold it there long enough for the thermometer to give an accurate reading.

TOILET INSTALLATION AND REPAIR

There are several reasons why you might need to remove a toilet. If you wish to install a new one, of course, you will first have to remove the old. You might also need to repair part of the bathroom floor or replace a broken cast-iron stool flange. If you intend to install new vinyl floor covering, you may wish to remove the toilet and reset it after the new flooring is down. Installing or resetting a toilet is not a difficult job, but it is one that contains a number of pitfalls. Learning to avoid or work around these pitfalls is at least half the job.

The basic components of a toilet

Labels: FLOAT ARM, FLOAT BALL, BOWL REFILL TUBE, TRIP LEVER, LIFT WIRES, OVERFLOW TUBE, GUIDE ROD, TANK BALL, FLUSH VALVE SEAT, BALLCOCK, HANDLE, TANK-FILL TUBE, GASKET, COUPLING NUT, SHUTOFF VALVE, TRAP, WAX GASKET, BOWL, RIM HOLES

The basic components of a toilet

Removing an existing toilet

While toilets are heavy and unwieldy, they are also fragile. To complicate the matter, old brass bolts and working parts become brittle with age and fall apart easily. If you wish only to remove a toilet in order to lay flooring, don't separate the tank from the bowl. While the unit will be heavier to move, you will avoid the possibility of damaging tank bolts and spud gasket seals.

Start by shutting off the water at the stool supply valve. Then flush the toilet and sponge the remaining water out of the tank. When the tank is dry, use a paper cup to dip the remaining water out of the bowl. Then, undo the water supply at the valve.

Next, pry the bolt caps from the base of the stool. Under these caps you will find the closet bolts. Older stools may have four bolts instead of the two bolts found on more modern stools. The front two bolts will be lag bolts with removable nuts and the back two will be standard closet bolts mounted through the stool flange. Use a small adjustable wrench to remove the nuts.

When you have the closet nuts loose and the supply tube disconnected, rock the bowl slightly in each direction to break it free from the floor and bowl gasket. Then grab the toilet by the bowl rim just in front of the tank and lift up. Because the bowl gasket will be sticky and dirty, have a newspaper ready to set the stool on. Then use a putty knife to scrape the remaining wax or putty from the stool horn and stool flange. Discard the old stool bolts. They are usually too damaged to be reused.

Resetting a stool

Before resetting a toilet, you will need to buy a new stool gasket and two new closet bolts. Stool gaskets come in two varieties. You can buy a traditional bowl wax or a rubber gasket. The advantages of a bowl wax are that one size fits all and that it costs much less than a rubber gasket. The disadvantage is that once used, a bowl wax cannot be used again. A rubber gasket can be used over and over again and tends to be more forgiving of beginner error.

Slide new closet bolts into the slots in the closet flange and center them so that each is the same distance away from the wall. Then press the stool gasket down on the flange. With these two items in place, you are ready to set the stool and fasten it down.

Lift the toilet as before, by the bowl rim near the tank, and carry it over to the flange. Carefully align the holes in the stool base with the closet bolts and slowly set the stool down. Press down evenly with all your weight. Then slide the cap retainers, closet washers and nuts onto the closet bolts and tighten each side a little at a time. When the base meets the floor and the nuts on the closet bolts seem snug, try rocking the bowl a little. If it moves, tighten the closet nuts another round. If the bowl does not move, stop. Both flanges and toilet bowls can break if you overtighten the bolts.

Use the stool a few days and then check the closet bolt nuts again. If the bowl has settled, tighten each nut another round or so until they feel snug. Then use a small hacksaw to saw the closet bolts off above the nuts. Finally, snap the bolt caps in place.

Clean away any wax

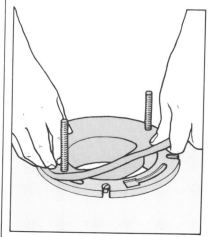

Press new gasket in place

Making the water connection

If you are resetting a toilet that you've just recently taken up, chances are you will be able to use the original stool supply tube. Apply a thin layer of pipe joint compound to the compression ferrule. Insert the compression end into the stool valve and then push down on the valve to gain the clearance needed to slide the cone washer end into the ballcock. If you cannot push the valve down, bend the supply tube slightly to gain the clearance you need. When both ends are in place, straighten the tube and tighten the compression nut.

If you add tile or another layer of flooring, you will not be able to use the original supply tube. The good news is that installing a new supply tube will probably be easier anyway, thanks to improved, more forgiving supply tube materials and designs.

While chromium-plated copper supply tubes are flexible and bendable, they kink very easily. And if the bend is too close to the end of the tube, the compression ferrule and the nuts will not slide over the pipe or seal properly. To avoid these installation problems, choose chromium-plated copper supplies that are ribbed. These more flexible supplies can almost be tied into knots without kinking. In tight situations, you can create a loop in the supply just to avoid cutting it to size.

Also on the market are plastic supply tubes encased in stainless-steel mesh. The plastic makes it flexible and the steel mesh makes it tough and durable. While this type of supply tube is more costly, it may be worth your while in time saved.

And finally, polybutylene plastic tubes are very flexible, inexpensive, corrosion-proof and easy to use. The problem with PB supply tubes, however, is that they are not universally accepted by code authorities. You will have to check with your local code administrators to see if they are allowable in your area.

No matter which type of supply tube you choose, you will find that the end that connects to the ballcock of your stool will be shaped to accept either a flat washer or a cone washer. The other end is designed to accept the compression nut and ferrule from the stool valve.

If the supply tube you intend to use must be cut to size, hold it against the ballcock threads and estimate how much you will have to bend it to make it meet the stool valve. Then make the two bends as near the ballcock end of the tube as possible. To avoid kinking the tube without using a tubing bender, apply steady even pressure at several points along the tube.

Hold the supply in place again and mark where the cut should be made. Be sure to add the depth of the valve socket to the length of the supply. Use a wheel cutter to make the cut. Then slide the ballcock nut on the tube from the unshaped end, followed by the compression nut and ferrule. Fit the compression end into the stool valve socket and the shaped end against the tapered ballcock and tighten the nuts. Tighten the compression nut one full round after you feel resistance. When tightening the ballcock nut, reach into the tank and hold the ballcock to keep it from spinning.

The procedure with other types of supplies is the same, except that with flexible supplies, you may not have to cut them to fit. PB supply tubes can be cut with a knife and come with their own plastic ferrules. Plastic supplies and ferrules should not be tightened as much as copper or brass—snug plus a half turn is usually plenty.

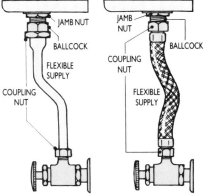

Chrome-plated tubing   **Mesh-encased polymer tubing**

SEE ALSO

Details for: ▷

| Toilet repairs | 30–34 |
| Tube cutter | 73 |

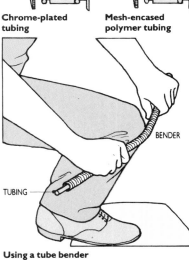

Using a tube bender

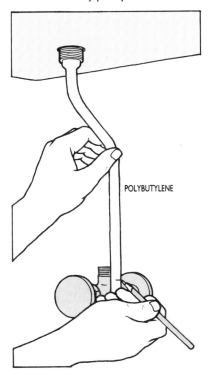

Mark and cut tubing

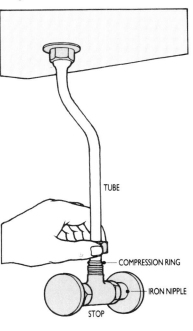

Connect to valve with compression fitting

SETTING A NEW STOOL

Your new toilet will come in two boxes. One will contain the bowl and the other the tank. You will have to install the bowl first, but open both boxes. Inside the tank box, you will find closet bolt caps and retainers needed to install the bowl.

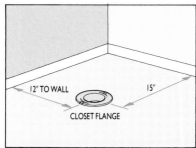

Center bolts on both sides of flange

WASHER AND NUT

SHIM

Shim level when floor is uneven

Typical tank assembly
1 Tank bolt
2 Rubber washer
3 Tank washer
4 Tank cushion
5 Spud nut
6 Spud washer

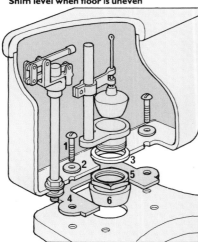

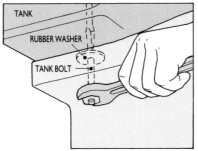

Tighten bolts until snug

Fitting a new toilet and tank

Insert new closet bolts into the slots of the stool flange so that both are the same distance from the wall. Then center the new bowl wax on the stool flange and press down to make even contact. With the bolts and wax in place, lift the bowl over the flange and guide the closet bolts through the holes in the stool base. If there is a slight angle in the position of the stool, straighten carefully before you press down on the stool. Then slide the bolt cap retainer rings and washers over the bolts and thread the nuts on. Tighten the nuts until you feel stronger resistance, push down on the bowl again and tighten the nuts another turn. Do not overtighten these nuts. It is always better to have to tighten them more after the stool has settled than to break the vitreous china bowl by overtightening the nuts.

Setting the tank
The same parts packet that contained the stool caps and retainer rings will contain a large rubber spud gasket and the tank bolts and washers. Press the spud gasket over the flush valve spud nut that protrudes through the bottom of the tank. Then sort the tank bolts and washers into sets. Start by sliding a

rubber washer onto each bolt and installing the bolts in the tank holes. Set the tank on the bowl so that the bolts go through the holes in the bowl and the tank rests on the spud gasket. Then install a brass washer and a nut on each bolt and tighten each bolt a little at a time to equalize the pressure.

The stool you buy may have two bolts or it may have three. The thing to remember is that the bolts should be tightened in sequence so that the tank rests evenly on the bowl when you are finished. Tighten them only until you feel firm resistance. With some models, the tank may not meet the bowl, but will remain suspended by the thickness of the spud gasket. After the tank is installed, connect the flush valve chain to the flush handle so that very little slack is left in the chain.

Making the water connection
Given the choice now available in closet supply tubes (◁), you are likely to choose one of the flexible types and avoid cutting and shaping chromium-plated copper tubes altogether. Flexible supplies cost a little more, but are much easier to use. This is an important consideration to many homeowners.

REPLACING A TOILET SEAT

The item most likely to wear out and often the most difficult to remove is your toilet seat. The nuts on metal seat bolts almost always rust or corrode to the bolts. When you attempt to loosen these nuts, the bolts, which are molded into the seat hinge, break loose inside the hinge and turn in place. The only alternative left is to saw the bolts off at bowl level with a small hacksaw.

To avoid chipping or scarring the porcelain surface, apply duct tape to the bowl around the seat hinge. In this way, you will be able to lay the saw blade flat on the bowl and cut under the seat hinge. It is tedious work, but once done it should not need to be done again. Seats manufactured today use plastic bolts and nuts.

Once you are rid of your old-fashioned seat with its brass bolts, your next replacement will be much easier. When shopping for a seat and lid, choose a painted wooden one. The plastic models on the market do not hold up as well.

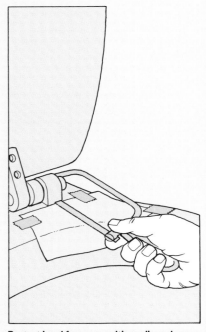

Protect bowl from saw with cardboard scrap

Toilet flanges do occasionally break. The method of repair depends mostly upon the material of the flange and soil pipe connected to it. Cast-iron flanges break most easily because of the nature of the metal. Brass flanges can also become brittle with age and break, while copper flanges will tear out at the slots. Plastic flanges can also tear or break at the slots. And, of course, any flange connected to a lead riser is easily threatened. When a toilet is tightened down too much, either the stool or the flange will break.

REPLACING A ROTTED FLOOR

When a stool is allowed to leak for months at a time, the water almost always damages the floor around it. If you have dry rot around your stool, you will have to take it up and replace a section of the floor. This is an involved task, but not a difficult one.

Take the stool up and cut the rotted layer of flooring out with a circular saw. Measure the area and cut a plywood replacement to fit the removed section. Then measure and cut the opening around the stool flange. Keep in mind that the flange rim must rest on top of the plywood.

Cut the plywood in two so that the center of the flange opening is the center of the cut. Slide each plywood half under the flange and nail down. Then screw the flange to the new floor.

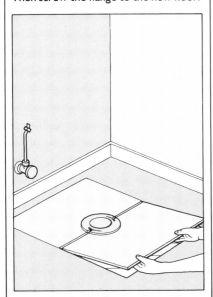
Slide plywood under flange

Replacing a defective toilet flange

If your cast-iron flange breaks at one of its side slots, you may be able to effect a quick-fix that is also permanent: You can buy a simple strap-metal repair item that works quite well in most situations. The strap is curved and shaped to slide under an existing flange. Just insert a closet bolt through the repair strap and slide the strap under the broken side of the flange. The strap is long enough to catch under the remaining edges of the flange. The pressure from the closet bolt keeps the repair piece tightly in place.

Replacing a cast-iron flange

If your cast-iron flange is badly broken, you will need to replace it. To remove a cast-iron stool flange, take up the stool and clean the excess putty or wax from the flange. If the flange is screwed to the floor, remove the screws. Then use an old screwdriver and hammer to pry the lead out of the joint between the flange and soil pipe. Hammer the screwdriver into the lead about ½ inch deep and pry up as you go.

Slide the new flange over the soil pipe and make sure that both closet bolt slots are the same distance from the wall. Then use heavy wood screws to fasten the flange to the floor.

To make a leakproof lead-and-oakum joint, push oakum into the joint and press it down so that it seats against the rim of the flange. Then use a hammer and packing iron to pack the oakum completely around the joint. Add more oakum until you have filled the joint two-thirds full with firmly packed oakum. Finally, fill the remainder of the joint with lead wool and pack it on top of the oakum until the hub is full.

Replacing a plastic stool flange

Usually one of the slots breaks or tears loose. If the ceiling below your stool is open, such as in a basement, the easiest way to remove a plastic stool flange is to cut the soil riser just below the flange. Then remove the flange screws and pull the damaged flange out.

Buy a new flange and a coupling. Then glue the new flange to a short stub of pipe and join the pipe to the riser with the coupling. Make sure the slots are the same distance from the wall before the glue sets (2). Then screw the flange to the floor and reset the stool.

If the soil riser is 4-inch plastic, cut the lip of the flange from above using a reciprocating saw. Saw around the joint where the top of the soil pipe meets the flange, then glue the flange inside the riser and screw the flange to the floor.

Removing a cast-iron flange and lead riser

Many older homes have cast-iron flanges connected to lead risers. Because lead is soft and becomes brittle with age, this combination should be replaced. Lead risers were used because lead is easier to work than cast-iron. One end of the riser would be bonded to a cast-iron insert that fits inside the nearest hub of cast-iron soil pipe. The other end comes through the bathroom floor and is flared out under a flat cast-iron flange. The best solution is to remove the flange and most of the lead riser and convert to plastic.

First remove the flange from the floor. Then cut the lead where it meets the cast-iron insert (**1**). Install a 4 × 3-inch no-hub coupling over the insert at the nearest cast-iron hub (**2**) and continue the rest of the way in plastic. The bolt slots should be 12 inches from the wall.

SEE ALSO

Details for: ▷

| Toilet repairs | 30–34 |

REPAIR STRAP
Broken flange

Glue flange to riser

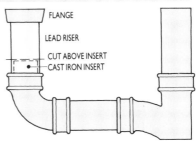
FLANGE
LEAD RISER
CUT ABOVE INSERT
CAST IRON INSERT

1 Connection details for installation with a lead riser

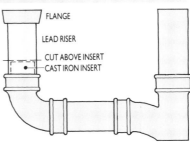
FLANGE
PLASTIC RISER
NO-HUB COUPLING

2 Connection details for installation with a plastic riser

TOILET REPAIRS

Toilets are a marvel of mechanical simplicity. With only a few moving parts, your toilet is responsible for half the water used in your home every day. It is easy to take toilets for granted, until they start to malfunction. Then we wonder how so few parts can cause so much trouble.

When you come to understand how a toilet really works, repair will no longer be a mystery. Mechanically, the components of a two-piece toilet haven't changed much since the turn of the century. The changes that have been made over the years have been modest. The changes you will see in the future will stem from the need to conserve water. You can expect toilets in the very near future to flush with under a gallon of water. Some European models work this way already. Even so, the age-old concept of gravity-flow flush valves and float-controlled ballcocks is likely to endure.

How toilets work

Your bathroom stool has only three mechanical components: the trip lever, flush valve and ballcock. The physical design of the toilet allows it to work so simply.

A two-piece toilet consists of a bowl, which rests on the floor, and a tank, which mounts on the bowl. The bowl contains a built-in trap that holds a consistent amount of water. The amount of water held by each model is determined by the height of the trap weir. The water trapped in the bowl keeps the bowl clean and keeps sewer gases from escaping from soil pipes into living areas.

The rim of the bowl is hollow. Water from the tank rushes into the rim and sprays through holes on its underside (and through a tube exiting opposite the drain opening). These holes are drilled at an angle, which causes the water to stream down the sides of the bowl at an angle. This angled spray serves two very important purposes. It scours and cleans the sides of the bowl, and it starts the water in the bowl spiraling more efficiently over the trap weir. The spiraling water also starts the siphoning action that pulls the water out of the bowl, over the trap and into the soil stack. Once the water in the bowl

begins siphoning over the trap, the water draining out of the tank keeps the siphon going until all wastes have passed into the soil pipe below. When the tank is empty and the flow is stopped, the siphon breaks, which causes the water climbing the trap to fall back into the bowl.

If you want to see a stool flush without a tank, pour a gallon of water directly into the bowl. The added water will force the trap, which will start the siphon, which will flush the stool.

The tank is a matter less of design and more of mechanics. All the tank does is hold water. Water is brought into the tank through a ballcock and is released through a flush valve. When you press down on the flush lever of your stool, a chain pulls a ball or flapper off the flush valve opening. Water then rushes into the hollow stool rim and spirals into the bowl. The water level drops in the tank, which lowers the ballcock float, which opens the ballcock to incoming water. When most of the water passes through the flush valve, the ball or flapper settles back into place and the tank begins to fill again. When the water level reaches a certain point, the float shuts off the incoming flow of water through the ballcock.

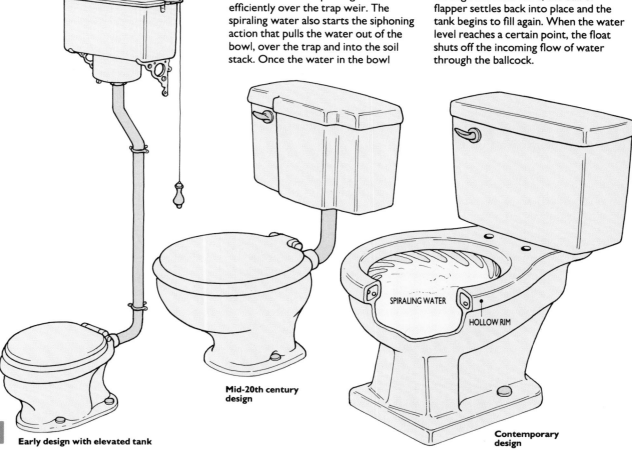

SPIRALING WATER

HOLLOW RIM

Early design with elevated tank

Mid-20th century design

Contemporary design

REPLACING A BALLCOCK SEAL

When a ballcock assembly wears out, some external component may break or the diaphragm gaskets may become too porous or brittle to make the seal. If an external part breaks, you will have to replace the entire ballcock assembly. If the internal seals wear out, you may wish to replace them without replacing the ballcock. The type of ballcock you have and its age will have a lot to do with the choice you make.

If the ballcock in your stool is made of brass and has been in service for many years, you should probably replace it. If your stool is not that old, or if the ballcock is made of plastic, you can probably get by with changing only the diaphragm seals (**1**). Start by shutting the water off below the stool or at the meter. Then flush the stool and undo the screws on the diaphragm cover. When all cover screws are removed, lift the float arm assembly and cover straight up and lay it aside. Remove all rubber or leather parts from the float arm assembly and diaphragm and examine the rim of the seat. If you can feel pits in the rim surface, you will have to replace the entire ballcock. If not, take the rubber washers and gaskets to your local plumbing outlet and buy replacement parts to match them.

Cover the new rubber washers and gaskets with heatproof grease and press them in place. On most models, you will have to replace the seat washer and the cover gasket.

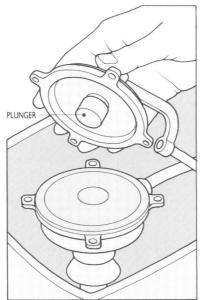

1 Replacing ballcock diaphragm seals

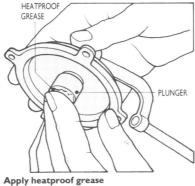

Apply heatproof grease

Replacing a ballcock

To install a replacement ballcock, shut the water off at the toilet or at the meter. Then flush the stool and sponge all remaining water out of the tank. Loosen the coupling nut that holds the supply tube to the ballcock inlet. With the supply tube nut off, remove the fastening nut (or jamb nut) just above it. While loosening the jamb nut, however, reach into the tank with your other hand and hold the top of the ballcock so that it does not spin into the overflow tube on the flush valve.

With the fastening nut removed, pull the ballcock straight up and out of the tank. If putty was used on the old ballcock, scrape any remaining putty from the bottom of the tank and clean the area with a rag.

To insure a leakproof connection, apply pipe-joint compound to the new ballcock gasket. Then insert the shank of the ballcock through the hole in the tank. From underneath the tank, thread the jamb nut onto the ballcock and tighten it with your fingers. Before tightening the nut completely, make sure that the float will not rub against the tank wall or catch on the flush valve overflow. Then tighten the jamb nut with a wrench until the gasket is flattened out and the nut feels tight. Fasten the refill tube inside of the overflow tube of the flush valve.

Connecting water to the new ballcock will pose one obvious problem. Because of the compression nut, the coupling nut will not slide off the supply tube, nor will the new coupling nut slide on. Either you will have to install a new supply tube or fasten the old nut to your new ballcock. If the old nut is in good condition, there is really no reason why you shouldn't use it, but a new supply tube and coupling nut will look more professional. Whether you install a new supply tube or not, be sure to coat the cone washer with pipe-joint compound or petroleum jelly before tightening the coupling nut.

Ballcock adjustments

While minor adjustments can be made by turning the screw on the float arm near the diaphragm, gross adjustments should be made by carefully bending the brass float rod. For a high water level, bend the float rod up. For a lower water level, bend the float rod down. Set the water level so that it is about 1 inch below the top of the overflow.

SEE ALSO

Details for: ▷
Adjusting ballcocks   32
Flush valves   32

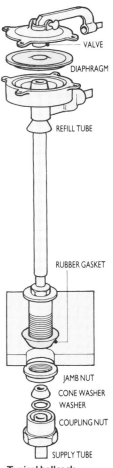

Typical ballcock assembly
All components are replaceable.

VALVE
DIAPHRAGM
REFILL TUBE
RUBBER GASKET
JAMB NUT
CONE WASHER
WASHER
COUPLING NUT
SUPPLY TUBE

CHOOSING A REPLACEMENT BALLCOCK

When replacing a ballcock, you will find that there are several designs on the market from which to choose. All work, but each has its advantages and disadvantages. Still widely used is the traditional brass ballcock with flush arm assembly. The term "ballcock" refers to this specific mechanism. The ball on the end of a float arm rises and falls with the water level, thus closing and opening a valve cock mechanism.

This design was universally used for so long that all mechanisms that fill a tank are often called ballcocks, even though their floats are not ball-shaped or their valves even float-operated. Brass ballcocks will last a long time, but they are considerably more expensive than some of their newer replacements.

The Fluidmaster design offers two

real advantages. First, Fluidmasters are available with antisiphon valves, which prevent contamination of household water, and second, they can be easily twisted apart at valve level should you need to clean the diaphragm. This second feature is particularly important if you live in an older home, where mineral deposits from an aging piping system frequently lodge under diaphragm washers.

A third design, marketed under the trademark Fillmaster, uses no float at all. Instead, a built-in regulator allows a measured amount of water into the tank and then shuts off automatically. The advantage of this mechanism is that it is simple. You can adjust the intake volume to meet the needs of the type of stool you have.

ADJUSTING BALLCOCKS

Adjusting a Fluidmaster valve

The only adjustment you will make on a Fluidmaster valve will be in the level of the float. To adjust the float, pinch the stainless-steel clip on the adjustment rod and slide the float up or down.

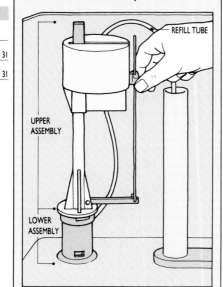

Fluidmaster fill valves

Adjusting a Fillmaster valve

The Fillmaster valve works without a float and relies on the hydraulic pressure of the water to open and shut the fill valve. An internal diaphragm measures the head pressure of the water to determine when to shut off. On top of the Fillmaster, you will see an adjustment screw marked "ADJ." Each complete turn of this screw will adjust the water level in the tank 1 inch. Fillmaster also offers an antisiphon model for cities that require them.

A Fillmaster uses no float mechanism

FLUSH VALVES

Flush valves can be very persistent sources of trouble. Luckily, most flush valve problems can be corrected with replacement tank balls.

Replacing a tank ball and assembly

Some toilets have a tank ball, trip wire and guide instead of a flapper and chain. If you need to replace just the tank ball, shut off the water to the toilet. Then reach into the tank, hold the ball still and thread the lift wire out of the top of the ball. If the ball is very old and brittle, the threaded inset may tear out of the rubber. If this happens, hold the inset with pliers and back the lift wire out.

With the tank ball removed, check the seat for calcium buildup and sand the valve seat if necessary. Then thread the lower lift wire into the new tank ball. You may also need to replace the top lift wires.

To replace a lift wire guide, use a thin-bladed screwdriver to loosen the guide from the overflow tube. Be very careful to avoid breaking the overflow tube off in the flush valve. If you do break the overflow tube, you will have to pry the remaining threads out of the flush valve with a pocket knife. Try to avoid damaging the threads inside the valve. Then coat the threads of the new tube with pipe-joint compound and carefully thread it in place.

Some flush valves have two lift wire guides clamped to the overflow tube, others require only one. In any case, fasten the guide clamps around the tube but do not screw them down tightly. Some adjustments will surely be necessary. Slide the lift wires through the guides so that they are centered over the flush valve. Then thread the lower lift wire into the new tank ball. Hold the upper lift wire next to the trip lever and move the wire and ball up and down to determine where the upper wire should fasten to the trip lever. Feed the upper wire through the nearest hole in the trip lever and bend it over on the other side.

Then, fill the tank with water and flush the toilet several times. The ball should drop straight down onto the flush valve. If it falls to one side, move the guides left or right to achieve the best alignment. When the ball drops straight on top the flush valve consistently, tighten the guide clamps and test several times more.

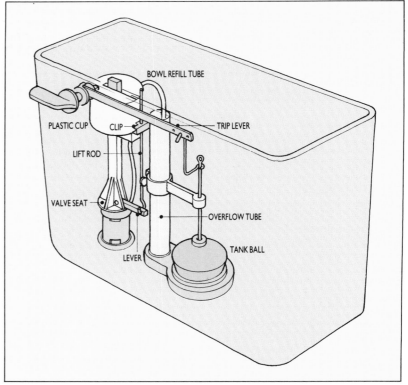

To install a tank ball, unscrew the ball from the lower lift wire

REPLACING A FLAPPER BALL

Flush valve problems occur when the flapper no longer holds water in the tank. In most cases, all that is needed is a new flapper.

To replace a flapper, first shut off the water to the tank. Then reach into the tank and carefully pull the rubber eyelets off the flush valve pegs. Disconnect the chain from the trip lever and discard the old flapper. Then run your finger around the flush valve to check for defects or pits in the rim of the seat. Even if no noticeable defects can be felt, it is a good idea to scour the seat rim with steel wool or emery cloth to remove any calcium buildup.

Some flush valves have no pegs through which the flapper eyelets hook. In this case, slide the collar of the flapper over the overflow tube until it seats against the bottom of the flush valve. Chances are that the flapper you buy will be designed for both applications and can be adapted to fit your needs. If your toilet's flush valve has no pegs, just slide the collar over the overflow tube without alteration. If the flush valve does have pegs, you will have to cut the collar from the flapper before hooking the eyelets to the pegs. Most universal replacement flappers are clearly marked so that you can make the cut in the right place.

After the flapper is in place, hook the chain to the trip lever so that there is not more than ½-inch slack in the chain.

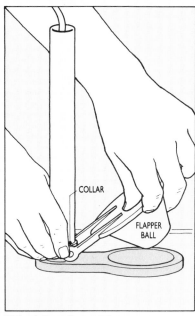

Unhook the collar eyelets

FAULTY FLUSH VALVES

Dealing with a faulty flush valve

If a flush valve seat is pitted or defective in any way, replacing the flapper or tank ball will do little good. The best solution is to separate the tank from the bowl and replace the entire flush valve. If this seems too intimidating, you might choose to install a replacement seat over the defective seat.

Installing a replacement seat

A seat replacement kit consists of a stainless-steel seat rim, flapper ball carriage, a flapper ball and epoxy putty. To install a seat replacement kit, you must first remove the flapper or tank ball and dry the defective flush valve completely. Press the new seat over the old so the epoxy is flattened evenly against both surfaces. Then allow the epoxy to dry for 12 hours.

Replacing a flush valve

A more permanent and much preferred solution is to replace the entire flush valve. To do this, you must first remove the tank from the bowl. Because tank bolts are likely to break in the process, you should buy new tank bolts when you buy the replacement flush valve.

Disconnect the water supply tube at the ballcock. Then remove the two or three tank bolts that hold the tank to the bowl. Lift the tank straight up and lay it on its side on the floor. Remove the large rubber spud washer from the flush valve shank. Then use a spud wrench or a large adjustable pliers to undo the spud nut from the old flush valve assembly.

If any putty or pipe-joint compound is stuck to the bottom of the tank, scrape it clean with a putty knife and sand the area around the opening. Then apply pipe-joint compound to the spud washer. Insert the spud through the tank opening and fasten it in place with the new spud nut. Make sure that the overflow tube is not in the way of the float arm inside the tank. Then press the large rubber spud washer over the spud nut and set the tank back on the bowl.

Apply pipe-joint compound to the rubber washers on the tank bolts and slide the bolts through the tank holes and bowl holes. (Some tanks require that you fasten the tank bolts to the tank with a second set of nuts and washers before setting the tank in place. Other models require rubber spacers between tank and bowl (▷).) Then tighten the tank bolts a little at a time until the tank rests firmly on the bowl, and reconnect the water supply tube.

Finally, attach the flapper ball to the overflow tube and make minor adjustments as needed. After the toilet has been flushed several times satisfactorily, check for tank leaks by running your hand under it and under the tank bolts.

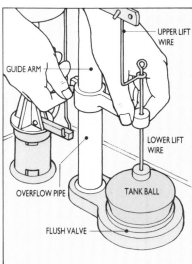

Adjusting a tank ball

Replacing a faulty trip lever

Replacing a handle/trip lever is not difficult, but the left-hand threads of the retaining nut have stumped a lot of beginner plumbers. The threads are machined on the shank counterclockwise so that the downward motion of the flush lever will not loosen the nut.

Take the lid off of the tank and use an adjustable wrench to loosen the retaining nut. When the nut is loose, slide it off the trip lever. (Some models have set screws instead of retaining nuts.) Then pull the lever out through the tank opening.

To install a new trip lever, feed the lever into the opening until the handle seats. Then slide the nut over the lever until it makes a right-angle turn and rests against the left-hand threads of the shank. Tighten the nut and connect the flapper chain to the most convenient hole in the lever. Follow with several test flushes.

SEE ALSO

Details for: ▷

Setting a new tank   28

Replacement seat kit
Kits come with epoxy putty.

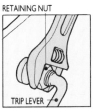

Replacing a tank trip lever
Loosen the set screw or undo the locknut.

33

MAINTAINING TOILETS

CLOGGED TOILETS

Occasionally, a toilet will clog and overflow. Often the water will seep away but the solids and paper will not. Most toilets clog at the top of the trap because that is where the trap is smallest. Clogs can be caused by too much paper or any of a number of bathroom items. Tooth paste caps, hair pins and combs are regular culprits.

Start by trying to plunge the toilet trap. A plunger with a collapsible funnel works best in toilets. If plunging does not do the job, dip the water out of the bowl and use a small pocket mirror and a flashlight to look up into the trap. If you can see the obstruction, chances are you can reach it with a wire hook.

If all else fails, rent a closet auger from your local tool rental outlet and crank the auger through the trap several times. A closet auger cable is just long enough to reach the toilet flange. When pulling the cable out of the trap, keep cranking in a clockwise direction to avoid losing the material causing the clog. Then test flush.

◁ Details for:
Drain-cleaning
techniques   13

SEE ALSO

Dealing with sweating tanks

Condensation appears on the outside of a stool tank when cold water from the water system meets the warm humid air of a bathroom. The water that collects on the surface of the tank eventually falls to the floor, often causing water damage over time.

Air-conditioned homes do not have tank sweating problems. If your home is not air-conditioned, your best alternative is to insulate the tank from the inside. There are polystyrene insulating liners on the market, but you can just as easily make your own, using ½-inch polystyrene or foam rubber.

Drain and dry the stool tank. Cut a piece for each wall and several pieces for the bottom of the tank. Then glue them in place with silicone cement and allow the glue to dry for a full day.

You can also reduce the temperature extremes by mixing hot water with the cold before it enters the tank valve. This method wastes a lot of hot water, of course, and unless you install a check valve on the hot water side of the connection, the other fixtures in your home will back-siphon hot water through their cold water outlets.

A temperature, or check valve, is available for this purpose and is not difficult to install. In a typical installation, you will tap into the hot water supply of your lavatory and run a ⅜-inch soft copper line to the temperature valve. The temperature valve should be installed just before the toilet valve. The cold water inlet of the temperature valve threads into the toilet valve and onto the toilet supply line, while the hot water pipe is connected to the temperature valve by means of a ⅜-inch compression fitting. If your basement ceiling is not finished, you can also tap onto the hot water line below the floor. A through-the-floor installation offers a much neater appearance. It also saves cutting into vanity cabinets.

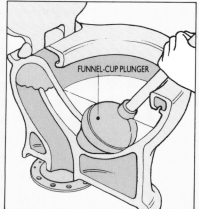

A funnel-cup plunger for a stool

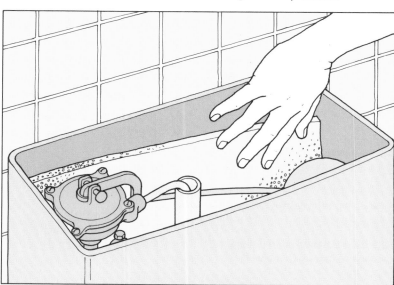

Glue foam rubber or polystyrene to the inside of the tank

Repairing wall-mounted stool tanks

Many older toilets have wall-mounted tanks that are joined to their bowls by means of a 90-degree pipe, called a "flush ell." While the working parts of these toilets are the same as in newer models, dealing with flush ells requires special care. When flush ells leak, carefully undo the nuts with a pipe wrench and wrap the threads with plastic pipe-joint sealant tape. When tightening the spud nut, hold on to the ell firmly to avoid cracking it.

Because wall-mounted tank stools waste so much water, and because their flush ells make them harder to repair, you should think about replacing them when they need extensive work. For minor repairs, choose repair kits that do not require taking off the flush valve. Instead of replacing flush valves in wall-mounted tanks, use one of the stainless-steel seat replacement kits.

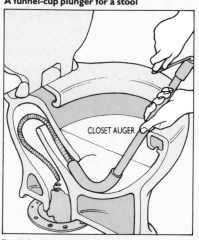

Feed the closet auger into the trap

Repairing a leaking flush ell
Loosen the nuts and wrap the threads with sealant tape.

BATHTUB INSTALLATION AND REPAIRS

For many years, bathtub designs consisted of various size free-standing leg tubs. Today, most tubs are of standard dimensions and are built into tub walls with an apron covering the side of the tub left exposed. Except for whirlpools, basic designs have changed very little in the last forty years, but some important material changes have taken place. Tubs today may be made of enameled cast-iron, porcelain-covered steel or molded fiberglass. The installation of tubs, however, remains much the same as it has always been.

Installing a built-in tub

To install a tub, you will need a framed opening that is 60¹/₁₆ inches long by at least 31 inches deep. On the drain opening side, cut a hole in the floor that is 8 inches wide by 12 inches long. Center this hole 15 inches from the back wall. Then nail braces between the studs all around the tub centered 14 inches above the floor.

Walk the tub toward the opening in an upright position. Stay to one side wall or the other. When you have the tub standing upright in the opening, gradually tip it down. At some point, the tub will likely turn on one corner and wedge itself diagonally in the opening. If your tub is made of fiberglass or steel, just lift and pull it into position. If your tub is a 375-pound cast-iron model, however, keep your feet and fingers away. Instead, use a 2-by-4 to pry the tub into place.

If your tub is made of steel or fiberglass, screw the lip of the tub to the backing you've installed between the studs. Cast-iron tubs have no lip and are held in place by subflooring and wall finish alone.

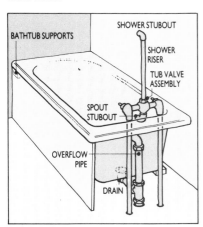

Typical tub installation

Installing a waste and overflow drain

The waste and overflow you buy will come in several pieces. Start by locating the drain shoe, drain gasket and drain strainer. Wrap a small roll of plumber's putty around the flange of the strainer. Then reach below the tub and hold the drain shoe against the tub opening with its rubber gasket sandwiched between the drain shoe and tub. Thread the strainer into the drain shoe.

Next, assemble the overflow tube, tripwaste tee and tailpiece, and connect the tee to the drain shoe with the compression nuts provided. You may also attach the tailpiece to the soil P-trap and ground joint adapter at this step in the process.

When the waste and overflow components are assembled and connected to the drainage system, you will be ready to install and adjust the tripwaste mechanism. Feed the plunger into the overflow tube and fasten the coverplate screws. Pour water into the tub and open and close the trip lever to determine what adjustments are needed for the best flow.

Adjusting a tripwaste

There are two basic tripwaste designs. One has a plunger cylinder attached to the end of the trip wire. The cylinder slides up and down inside the tripwaste tee when you activate the trip lever. In the down position, the cylinder slides into the tee and closes it.

The other design features a pop-up lever and plug in the drain opening of the strainer. This model has a large spring attached to the lift wire that moves the pop-up lever up or down. When the spring is in the down position, the pop-up lever pushes the plug up and drains the tub. When the spring is in the up position, the pop-up seats itself in the drain opening and closes the drain.

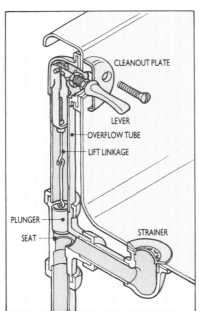

Typical plunger-type waste and overflow

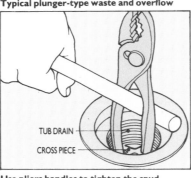

Use pliers handles to tighten the spud

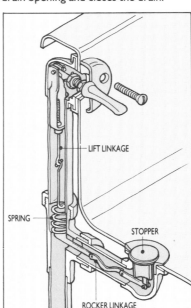

Typical pop-up waste and overflow

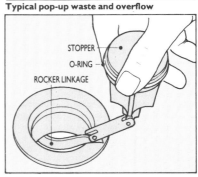

Pull the pop-up plug and lever out

SEE ALSO

Details for: ▷
Showers   37–38

Adjusting a tripwaste
Loosen the locknut on the lift wire and thread the wire up or down.

35

MAINTAINING BATHTUBS

Tub valves and shower heads

Most tub/shower valves on the market come with a diverter spout and shower head. Start by installing the faucet body. Bring ½-inch copper water supplies up in the tub wall to a height of 28 inches. Then cut a 44-inch length of copper for the shower riser and a 4½-inch length for the tub spout rough-in. Thread or solder the valve in place, close enough to the wall so that the coverplate screws will reach through the tile and drywall and into the faucet. Then solder a sweat/FIP fitting, called a "drop eared ell," to the shower stem and to the spout leg. Plug the shower head fitting and the spout fitting and turn the water on to test the installation. Turn both the hot and cold sides of the valve on and watch for leaks.

After testing the piping for leaks, drywall and tile the walls and thread the shower head into its fitting with plastic pipe-joint sealant tape. To install the spout, measure from the surface of the tile to ⅜ inch inside the spout fitting and buy a ½-inch nipple that length. Wrap sealant tape around both ends of the nipple and thread the nipple into the spout fitting. Then turn the spout onto the nipple and caulk around the spout and faucet coverplate.

SEE ALSO

◁ Details for:
Drain-cleaning
techniques   13
Showers   37–38

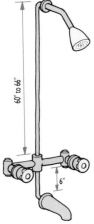

Installing a shower
Rough-in a tub valve
28 in. from the floor.

Push grout into joints

Maintaining tiled tub walls

Eventually, every tiled tub/shower wall will need repair, but you can extend the life of your tile with a few simple maintenance procedures. The danger signs are loose or missing grout and excessive mildew in grout joints. To regrout a tiled wall, start by digging all soft or loose grout from between tiles with a grout removal tool, available from any tile outlet. All joints to be regrouted should be scraped to a depth of at least 1/16 inch. Then wipe away the loose grout and clean the entire wall with a good tile cleaner.

With the tile prepared, select a small container of premixed, ready-to-use grout and force a liberal amount into each prepared joint with your fingers. Use a damp sponge to smooth the grout. Wipe in large diagonal patterns until the grout is uniform. Then allow it to set for one-half hour and wipe the surface again to remove any residue. After the grout has cured for 24 hours, apply clear silicone sealer to the entire wall with a soft cloth. Reseal your tile at least every six months thereafter.

Replacing ceramic tiles

When water is allowed to seep behind tiles, it ruins the tile mastic and the wallboard. Eventually, tiles will loosen and fall out. To replace them, you may need to remove all tiles that have come in contact with moisture and replace a section of drywall.

Use a knife or screwdriver to pry under the tiles. If tiles come up easily, take them out. Then cut out the affected wallboard and nail a new piece of water-resistant wallboard in its place. If the edges fit together neatly, you will not have to tape the seams. Prime the new wallboard with clear sealer or oil-base paint and allow the primer to dry completely.

To strip the paper and mastic from the removed tile, soak each tile in very hot water and scrape it clean with a putty knife. Lay the tiles out on the floor in the order in which they will go back on. Apply wall-grade tile mastic to the wall or tile with a notched trowel. A ⅛-inch notch will provide enough gap in the cement to hold the tile to the wall.

With the cement on the wall, press the old tiles back in place. Clean away any tile cement from the tile surfaces and allow the cement to cure for 48 hours. (If you grout the joints too soon, the gases escaping from the cement will cause pinholes to appear in the grout.) When the cement has cured, grout the new tile joints and seal the entire wall with clear silicone sealer.

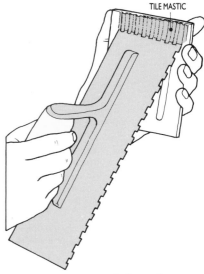

TILE MASTIC

Apply cement with a notched trowel

Adjusting and cleaning a tub drain

Often slow-draining tripwastes just need adjustment. Start by undoing the screws that hold the coverplate to the overflow opening. Pull the coverplate and lift wire up and out of the overflow tube. Part way down the wire, you will see adjustment slots, or a threaded adjustment wire. If your lift wire has a slot adjustment, pinch the two bottom wires together and move them up or down into the next slot level. If your tripwaste has a threaded wire adjustment, loosen the locknut and thread the wire up or down about ⅛ inch and retighten the locknut. Then slide the tripwaste back into the overflow and replace the coverplate screws.

If you do not find a clog at the tripwaste mechanism, you will have to snake the trap and drain line. Because tubs are snaked through the overflow and not through the drain, you will need to remove the coverplate and tripwaste components. Feed a small hand snake into the overflow until you feel resistance at the trap bend. When you feel the trap, start cranking the snake in a clockwise direction while pushing the cable slightly. After you crank through the trap, pull the snake out to see if you've snagged the clog. If not, snake the entire tub line to where it enters the main stack. Then replace the tripwaste components and flush with plenty of hot water.

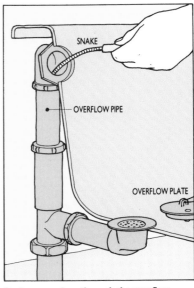

SNAKE

OVERFLOW PIPE

OVERFLOW PLATE

Snake away clogs through the overflow

Shower stalls come in several varieties. All-metal or plastic freestanding showers can be installed anywhere near a floor drain. They are popular in unfinished basements and are often thought of as utility showers.

One-piece fiberglass stalls are built into framed walls and are popular in finished bathrooms, both upstairs and down. They *drain into dedicated traps and are plumbed permanently.*

A more traditional shower consists of a separate pan built into a framed wall and plumbed into a dedicated trap. The framed walls are covered with water-resistant drywall or concrete board and finished with ceramic tile, molded plastic or fiberglass shower walls.

Installing a freestanding shower

A freestanding shower consists of a raised pan, three wall panels, cornerbraces, a drain spud and a valve and shower head. All of these parts will come in a box and must be assembled on site. Start by setting up the pan and plumbing the drainpipe to the nearest floor drain. Then install the walls and cornerbraces and fasten the walls to the pan according to directions.

Next, assemble the valve and shower riser and mount the valve and shower head to the plumbing wall of the shower. Some freestanding shower stalls will have predrilled valve holes, but other models will have to be drilled. If you are plumbing the supply lines in steel, use a union on each side. If plumbing in plastic or copper, plumb into the valve with male adapters.

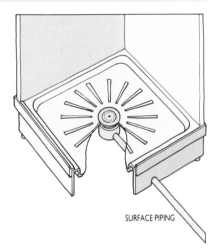

SURFACE PIPING

Plumbing the drain pipe
Run surface piping to the nearest floor drain.

A freestanding shower stall

Installing a one-piece fiberglass shower

To install a one-piece fiberglass shower, start by framing the walls. The width of the opening should not be more than $1/16$ inch wider than the width of the fiberglass stall. The depth of the opening should exceed the front drywall lip of the shower by 2 inches. With the framing completed, cut the drain hole in the floor. Take the measurements from the bottom of the shower stall to find the center of the drain, and make the opening at least 5 inches in diameter so you will have room to work.

Next, install the drain spud in the pan opening. Wrap the underside of the drain flange with plumber's putty and press it into the opening. Then slide the gasket over the spud from below and tighten the spud nut.

With the drain in place, you will be ready to set the shower in its frame. Because fiberglass shower floors tend to flex when you stand in them, it is a good idea to support the floor with a little perlited plaster. Mix enough wet plaster to cover the wooden floor 1 inch deep and about 1 ½ feet around the drain hole. Then set the shower in place on

top of the plaster. Level the shower walls and step into the shower to settle the base into the plaster. Nail the wall-board lip of the shower walls to the studs with galvanized roofing nails and connect the drain spud to the P-trap below the floor.

With the stall in place, measure for the shower valve cut. The valve should be 48 inches off the floor and the shower head should be 6 feet or more from the floor. Use a hole saw to cut the shower valve holes. Solder the shower valve, supply lines and shower head together and mount the shower valve through the stall wall. With the valve in place, install the shower head and nail 2-by-4 braces behind the valve and the shower head. Secure the valve and head firmly to the braces with pipe hooks. Then connect the water supply lines to the water system and test the solder joints under full pressure.

FLANGE   SHIELD

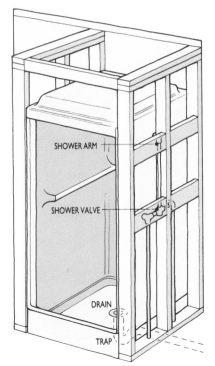

SHOWER ARM

SHOWER VALVE

DRAIN

TRAP

Framing the opening
Frame an opening the exact size of the shower.

Hardboard shield
Use hardboard to keep from hitting the shower wall.

INSTALLING SHOWERS

Installing a fiberglass shower

Like a one-piece shower, a shower pan is installed in a framed opening. Simply frame the stall as you would with a one-piece shower and cut the drain opening in the floor. Install the drain spud in the pan, using putty under the flange, and set the pan in place.

Shower pans often have metal drain spuds. The drainpipe from the P-trap then extends through the spud to just below the drain screen. You can use a rubber gasket to seal the joint or use a traditional lead wool and oakum seal.

With the pan installed and connected to the drain line, install the valve and shower head rough-in in the framed wall and secure both with pipe hooks and 2-by-4 bracing. Test the piping and cover the walls with water-resistant wallboard. You can then cover the wallboard with ceramic tile or a molded shower surround.

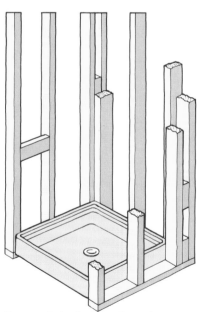

Frame walls for the pan and trap riser

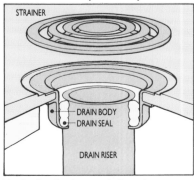

STRAINER

DRAIN BODY
DRAIN SEAL

DRAIN RISER

Use lead-and-oakum or drain gasket

Installing a tub or shower surround

Molded fiberglass or plastic tub and shower surrounds are easy to install and offer long-term durability. The appeal of these molded shower walls is that they have very few seams, and therefore few opportunities to leak. The only situation that prohibits the use of surrounds is crooked walls. Even with slightly out-of-plumb walls, a little bottom edge trimming will create an effective seal.

Most tub or shower surrounds come in three pieces. You will have to cut the valve and spout rough-in holes, but beyond that, they are ready to install. Before installing any of the panels, put a level on all walls and on the top of the tub to make sure they are reasonably plumb and level. If everything is straight and level, mark the exact center of the back wall of the tub. Then mark the center of the back wall shower panel.

Apply several beads of panel adhesive to the back of the center panel. Then peel the paper from the adhesive strips around the edges, if present. Lift the panel up to the back tub rim so that the bottom of the panel is an inch away from the wall. Rest the panel on several match sticks laid on the back rim of the tub. When the center of the panel is aligned with the center of the back wall, press the bottom of the panel against the wall. Work from bottom to top until the adhesive strip has sealed the entire panel. Then rub the panel firmly with the palm of your hand to flatten the panel adhesive to the wall.

Next, install the corner panel opposite the plumbing wall. Use the same method you used for the back panel, but press the corner in first. The corner panel will lap the back panel by several inches.

To cut the valve handles and tub spout in the plumbing wall panel, remove the spout and handles and measure from the tub rim and inside corner. Use a hole saw to cut the openings. Even a small cutting error will ruin the panel, so double-check all measurements before cutting. With the holes made, slide the panel over the handle stems and spout rough-in to make sure that everything fits. Then apply panel adhesive, peel the paper from the adhesive strip and press the corner in place.

When the adhesive dries, caulk the bottom seam and the valve flanges with white silicone sealant and both corners with latex tub and tile caulk.

Measure from the inside wall to the center

Set bottom of wall panel first

Apply panel adhesive liberally

It is hard to imagine living an active life without instant access to hot water. When your water heater fails, don't panic. A little troubleshooting, adjustment and repair may extend its life.

Water heaters are fairly simple appliances, but when problems arise, they can present an array of confusing symptoms. Because problems can occur in any part of your hot-water system, don't limit your investigation to the heater alone. The diagnostic charts on the next page will help you locate the source of your hot water trouble.

The piping system

In some cases, water heater problems turn out to be piping problems instead. For example, high operating costs can often be traced to dripping faucets or leaking pipes. Several dripping faucets in your home can waste hundreds of gallons of water a year. A simple, inexpensive faucet repair can pay for itself in energy saved.

Long, uninsulated piping runs also waste hot water. When you draw water from a faucet at the end of a run, hot water from the tank must first push the cooled water through the pipe. This not only wastes water, but energy as well. Uninsulated pipes dissipate heat much as a radiator does. To keep the energy you buy from escaping through the walls of hot water pipes, you should consider insulating all hot water lines.

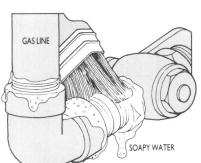

Test for gas leaks with soap and water

Problems inside the tank

An aging water system can carry *sediment* into a tank, or sediment may collect in flakes of calcium and lime. In electric models, sediment-covered heating elements will burn out quickly. In gas water heaters, sediment accumulates in the bottom of the tank and forms a barrier between the heat source and the water. Not only does sediment make your heater very inefficient, but air bubbles created by the heat percolate through the sediment and cause a continuous rumbling sound. So, if your electric heater burns up lower elements frequently, or if your gas heater rumbles, sediment may be the culprit.

To remove sediment, drain as much water as possible from the tank. Then turn the water supply on and allow the new water to flush through the drain valve for a few minutes.

Dip tubes

A *dip tube* is a plastic pipe that delivers incoming cold water to the heat source near the tank bottom. Occasionally, a dip tube will slip through the cold-water inlet fitting and fall into the tank. When this happens, cold water entering the tank is drawn through the hot-water outlet without being heated. To replace a dip tube, disconnect the inlet pipe from the tank. Cut a length of ½-inch I.D. soft copper pipe long enough to reach within 12 inches of the tank bottom. Flare the pipe end so it is large enough to rest on the rim of the inlet fitting. Slide the tube into the fitting and reconnect the inlet pipe.

Anode rods

New water heaters are equipped with magnesium *anode rods* that coat any voids that develop in the porcelain tank lining. An anode rod acts as a sacrificial element. It has a slightly different electrical charge than the other metals in the tank and draws rust and corrosion to it. These rods are usually trouble free, but problems can occur when water contains an unusually high concentration of dissolved mineral salts. As a result, the water will have a gassy odor or taste. To correct this problem, unscrew the magnesium rod and replace it with an aluminum rod.

Relief valves

A *relief valve* keeps a heater from exploding in the event a thermostat becomes stuck. When pressure builds and the water gets too hot, the relief valve opens until the pressure is equalized. However, the spring mechanism in some valves weakens with age and valves release water with the slightest variation in pressure. To correct this, simply remove the old valve with a pipe wrench and thread in a new one.

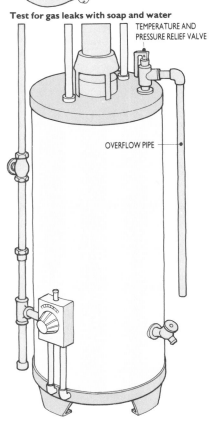

Testing valves Water should rush out if not stuck.

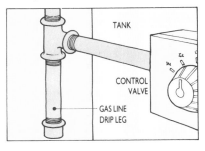

Drip legs
Install a drip leg near the control valve.

GAS AND ELECTRIC WATER HEATERS: SYMPTOMS

GAS WATER HEATER DIAGNOSTIC CHART

SEE ALSO

◁ Details for:
Common water
heater problems   41–42

CAUSES	Burner will not light	Burner flame floats—Lifts off	Burner flame yellow—Lazy	Burner flame noisy	Burner flame too high	Burner pops when turned off or on	Flame burns at orifice	Pilot will not stay lit	High operating costs	Insufficient hot water	Slow hot-water recovery	Pounding and steaming at faucet	Dripping relief valve	Thermostat fails to close	Condensation	Combustion odors	Smoking—carbon formation	Pilot flame too small	Pilot flame too large	SOLUTIONS
Insufficient secondary air		•		•						•						•	•			Provide ventilation
Dirt in main burner orifice	•		•		•	•		•	•	•						•	•	•		Clean—Install dirt trap
Dirt in pilot burner orifice							•												•	Clean—Install dirt trap
Flue clogged		•	•				•	•		•					•	•	•			Remove—Blow clean—Reinstall
Pilot line clogged	•						•											•		Clean—Install dirt trap
Burner line clogged	•		•				•													Clean—Check source and correct
Wrong pilot burner	•						•											•	•	Replace with correct pilot burner
Loose thermocouple							•													Finger-tight plus ¼ turn
Defective thermocouple lead	•						•													Replace thermocouple
Defective thermostat	•				•					•	•									Replace thermostat—(Call plumber)
Improper calibration									•	•	•	•	•	•						Replace—(Call plumber)
Heater in confined area	•	•	•												•	•	•			Install vent in wall or door
Heater not connected to flue		•	•	•											•	•	•			Provide and connect to proper flue
Sediment or lime in tank									•	•	•	•								Drain and flush—Repeat
Heater too small									•	•	•									Upgrade to larger heater
Gas leaks								•												Check with utility—Repair immediately
Excess draft		•		•				•		•										Check source, stop draft
Long runs of exposed piping									•	•										Insulate hot lines only
Surge from washer solenoid valve													•							Install air cushion pipe
Faulty relief valve													•							Install rated T & P valve—Soon
Dip tube broken									•	•	•									Replace dip tube

ELECTRIC WATER HEATER DIAGNOSTIC CHART

CAUSES	No hot water	Insufficient hot water	Slow hot-water recovery	Steaming and pounding at faucet	High operating costs	Dripping relief valve	Excessive relief valve operation	Condensation	Element failure	Blown fuse—Tripped circuit breaker	Service wires charred or hot	Continuous operation	Singing thermostat	Wet heater insulation	Gas odor or taste in water	Fluctuating temperatures	Rusty or discolored water	Rumbling-pounding in tank	SOLUTIONS
No power	•									•									Check fuses breakers—Reset
Undersize heater		•		•								•			•				Install larger heater
Undersize elements		•	•									•							Replace with rated element
Wrong wiring connections	•	•		•					•	•	•								See manufacturer's instructions
No relief valve			•																Install relief valve—Soon
Leaking faucets		•		•									•						Locate and repair
Leaks around heating elements	•			•		•						•							Tighten tank flange
Sediment or lime in tank		•															•	•	Drain and flush—Water treatment?
Lime formation on elements		•	•														•	•	Replace elements
Thermostat not flush with tank		•	•	•	•						•	•		•					Reposition
Faulty wiring connection	•	•	•					•	•		•			•					Locate, reconnect
Faulty ground		•	•						•					•					See maker's grounding instructions
Short	•			•	•				•	•	•								Locate short circuit—Correct
Gas from magnesium anode rod															•		•		Install aluminum anode rod
Damage from electrolysis																	•		Install dielectric unions
Excessive mineral deposits			•														•		Flush tank—Install water filter
Improper calibration	•				•	•	•	•				•							Replace thermostat—(Call plumber)
Eroded anode rod															•	•			Replace
Faulty thermostat	•	•	•	•	•	•						•			•				Replace—(Call plumber)
Faulty high limit (ECO)	•	•	•								•			•			•		Replace
Open high limit (ECO)	•	•																	Reset button or replace
Dip tube broken		•	•		•							•			•				Replace dip tube

COMMON PROBLEMS WITH GAS WATER HEATERS

A typical gas heater consists of a steel tank, a layer of insulation and a sheet-metal jacket.

The bottom of the tank is heated by a gas-fixed burner that is controlled by a thermocouple and a regulator valve.

To vent excess heat and noxious fumes, a gas heater tank is equipped with a hollow tube through its center, which connects to a home's flue.

A supply of secondary air

For a gas heater to burn evenly and efficiently, it must have an ample supply of *secondary air*. If your water heater shares space with a furnace and clothes dryer, then a continuous air supply is especially important, because they compete with the heater for air. When a heater is starved for air, the flame will burn orange, jump and pop. An orange flame means higher operating costs. Be sure that the heater has a sufficient supply of secondary air by opening doors in confined areas or by installing louvered vents in the doors.

A clogged flue

A *clogged flue* is caused by rust or debris that accumulates at tight bends in the flue piping. A clogged flue is a serious health hazard. Deadly carbon gases, unable to vent through the roof, are forced into living quarters. An easy way to check if the flue is working properly is to place a lit cigarette near the flue hat while the heater is on. The smoke should be drawn into the flue. To locate an obstruction, turn the heater to pilot, disassemble the tin vent pipes and inspect and clean each piece, then reassemble the flue.

Dirt in gas lines

Dirt in gas lines often makes its way into the heater's controls and burner systems. A dirty pilot line or burner line will cause the heater to burn unevenly or to stop burning entirely. To clean these lines, disconnect them from the regulator and slide a thin wire through each line. Then, blow air through the lines. If dirt is lodged in the gas control valve, call a plumber. Control valves are delicate mechanisms that can be dangerous if serviced improperly.

Thermocouple breakdown

A *thermocouple* is a thick copper wire that has a heat sensor on one end and a plug on the other. Heat from the pilot flame sends a tiny millivolt charge through the wire, which causes the plug to open the control valve. When a thermocouple's sensor burns out, the heater's magnetic safety valve remains closed and the pilot light won't burn.

To replace a thermocouple, turn off the gas and disconnect the entire burner assembly from the control valve. Remove the thermocouple from its retainer clip near the pilot and snap a new one in its place. Be sure to position the sensor directly in line with the pilot flame. Finally, reconnect the burner assembly to the control valve.

Gas leaks

If you smell a strong gas odor, it's likely there is a *gas leak*. Leave the house immediately and call your gas or utility company. If you smell only a slight trace of gas, it may be a leaky pipe joint. To find the leak, brush every joint with a mixture of dish detergent and warm water. Soap bubbles will appear around the leaky joint. Shut off the gas at the meter. Bleed the line at the union located above the heater and ventilate the area.

Cut the pipe a few inches away from the leaking fitting. Unscrew the bad fitting and thread a new fitting in its place. Then thread the cut pipe end. Connect the new fitting to the newly threaded pipe with a short nipple and union. Finally, turn the gas back on, bleed the air from the line and retest all pipe joints with soap and warm water.

SEE ALSO

Details for: ▷
Electric water heater problems	42

Typical gas water heater
1 Tin vent
2 Cold water inlet
3 Hot water outlet
4 Flue hat
5 Union
6 Relief valve
7 Discharge pipe
8 Anode rod
9 Water
10 Tank
11 Dip tube
12 Insulation
13 Flue baffle
14 Gas control
15 Gas pipe
16 Temperature control
17 Gas valve
18 Burner
19 Draincock
20 Thermocouple lead
21 Pilot line
22 Burner supply
23 Thermocouple

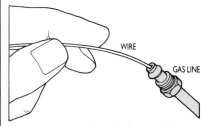
Use a thin wire to clear dirt from gas lines

WIRE
GAS LINE

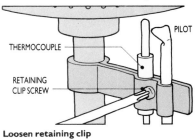
Loosen retaining clip

THERMOCOUPLE
PILOT
RETAINING CLIP SCREW

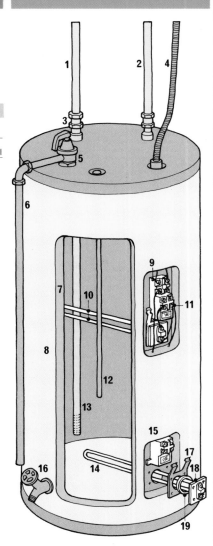

**Typical electric
water heater**
1 Inlet
2 Outlet
3 Union
4 Power cable
5 Relief valve
6 Discharge pipe
7 Insulation
8 Tank
9 High limit
10 Upper element
11 Upper thermostat
12 Anode rod
13 Dip tube
14 Lower element
15 Lower thermostat
16 Draincock
17 Bracket
18 Element flange
19 Gasket

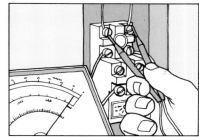

Test switch with ohmmeter

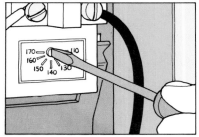

Adjust the new temperature setting

COMMON PROBLEMS WITH ELECTRIC HEATERS

If your electric water heater fails, first check for burned-out fuses or tripped circuit breakers at the main service panel. If the problem is not in the service panel, go to the heater. Remove the access panels and press the reset button on each thermostat and listen for a ticking noise caused by expansion as the elements begin to heat up. If this procedure doesn't produce hot water, the problem may be in the wiring, thermostats or elements.

Loose wires

Check to be sure that no wire has come loose from its terminal. If a wire is loose or disconnected, turn off the power to the heater before refastening the wire.

Defective thermostat element

To determine if the problem is in the element, thermostat or high-limit protector, test each part with a volt-ohmmeter (VOM). If you do not have a VOM, try simple logic. If the heater produces plenty of warm water but no hot water, then the top element or thermostat is probably defective. If you get a few gallons of very hot water followed by cool water, then the bottom element or thermostat probably needs replacing. Since elements fail much more often than thermostats, assume a faulty element or test with a VOM.

Replacing an element
To replace a defective element, first shut off the power and water supply to the heater. Next, drain the tank to a level below the element to be replaced. Disconnect the wires from the terminals and unscrew the element. Pull the element straight out of the tank. Then clean the gasket surface, coat it lightly with pipe-joint compound and seat a new gasket. Attach the new element to the heater and reconnect the wires to the terminals. (Some elements thread into a threaded tank opening, while others bolt to a gasket flange.) Before turning the power on, fill the tank with water and bleed all trapped air through the faucets. An element that is energized when dry will burn out in seconds.

Finally, replace the insulation, thermostat protection plates and access panel. Then, turn on the power. If after 45 minutes you still don't have sufficient hot water, then a replacement thermostat is in order.

Replacing a thermostat
Shut off the power and disconnect the wires from the thermostat's terminals. Pry out the old thermostat and snap the new one into the clip. Then, reconnect the wires, replace the insulation and turn the power back on. Allow both elements to complete their heating cycles and then test the water temperature at the faucets using a meat thermometer. Adjust the thermostat until the water temperature is between 130 and 140 degrees F.

New designs in electric water heaters

For years, electric water heaters have been made with metal storage tanks. All other components were replaceable, but when a tank developed a leak, the entire heater had to be replaced. The longevity of the tank, then, determined the longevity of the heater. While most manufacturers still prefer metal tanks, at least one offers a plastic tank. Because plastic cannot rust through, and because mineral salts will not adhere to it, this new design seems to have real potential.

Another recent design rejects the principle of storing hot water entirely.

The makers of this design maintain that heating and reheating stored water is too wasteful. They offer, instead, a system that heats cold water as it passes through a heating element. In this way, only the water used is heated. With careful use and planning, these units should offer real savings. If you regularly take showers while your clothes washer or dishwasher are operating, then this system may have trouble keeping up. In any case, consider your needs before investing.

WATER SOFTENERS AND WATER FILTERS

The water we pump into our homes varies greatly in quality from region to region, and even from well to well. The degree of mineral content in groundwater accounts for these differences and can also account for a few health and plumbing problems as well. Most municipal water systems *provide water that falls within tolerable limits of hardness and dissolved mineral salts. Others, especially rural systems, do not. When mineral levels are too high, water must be treated or filtered to bring it within the acceptable tolerances for domestic use.*

Water softeners

The purpose of a water softener is to substitute sodium for calcium, magnesium or iron. These minerals, in high enough concentrations, can cause clogged pipes and appliances and can give drinking water a foul smell. Water softeners neutralize these minerals, which makes conditioned water feel softer. It helps eliminate soap scum on fixtures and reduces the amount of mineral sediment in water heaters.

Water softeners cause problems of their own, however. They naturally raise the salt content of drinking water. Too much salt presents certain health risks, especially for those on low sodium diets. There are two sides to this issue, of course, but softeners are probably installed more often than are needed. If soap does not dissolve well in your water or if mineral buildup occurs on your fixtures, then you may need a softener. If you are uncertain, you might have your water tested by a local lab or by your state health department.

If you do need a water softener, make sure you isolate your main cold water drinking faucets and your outdoor hydrants. Salt water will kill a lawn in short order. The easiest installation in a home with finished basement ceilings is to tie the intake of the softener to the inlet line of your water heater. With a hot-water-only installation, you get soft water where you need it most, in your clothes washer and dishwasher. You will also get some soft water in your tub/shower or wherever you mix hot and cold water. You will also prevent mineral buildup in your water heater.

For a more complete installation, you should tie the softener into the incoming water trunk line before it reaches any branch fittings. To isolate cold water drinking faucets and outdoor hydrants, cut and cap the branch fittings that serve these lines and tie all hard water lines in at a new location, somewhere between the meter and the soft water inlet fitting.

All water softeners must be equipped with a three-way bypass valve or a three-valve bypass configuration. You can buy one three-way valve and splice it into the inlet and outlet lines of the softener, or you can install a separate globe valve in the inlet line and another in the outlet line. Then install two tees in each pipe above these valves, joined by a third valve that will act as a bypass. When the softener is in service, the two line valves are open and the center valve is closed. When the softener is not in service, the two line valves are closed and the center valve is opened to allow water to pass from the inlet line to the outlet line without entering the water softener.

Water filters

Unlike softeners, which treat mineral salts, filters only screen unwanted sediment, minerals and chemicals. They strain water through fiber or charcoal cores designed for specific problem water conditions. While most filters are used to trap sediment, others are specially designed to filter mineral salts, like iron, calcium and sulfur. Other filter cores are designed to trap health-threatening nitrates, although the effectiveness of nitrate filters is debated by experts.

Many of these filters can be reused after they are back-flushed, but others must be discarded when they reach a saturation point. In any case, you should research filters thoroughly before investing in a filter system.

SEE ALSO

Details for: ▷
Typical plumbing
system   4

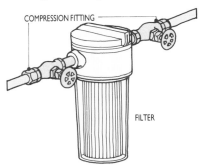

Horizontal installation

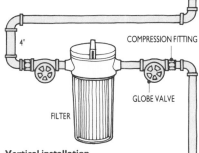

Vertical installation

How to install a filter

Filters are easy to install, but keep in mind that they must remain upright to work. Horizontal installations are easiest. Simply cut a section of pipe from the main trunk line and install gate valves on each new end. Then install the filter between the two valves with nipples and unions. With copper or plastic pipe, use male adapters.

For a vertical pipe installation, you will have to cut the pipe and create a horizontal loop for the filter. Use gate valves unions and nipples to make the necessary connections.

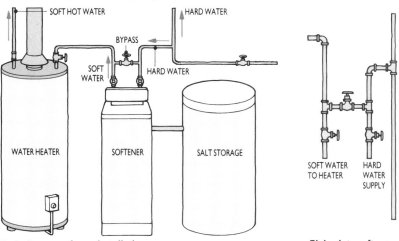

Typical water softener installation   **Piping into softener**

RURAL SEPTIC SYSTEM MAINTENANCE

Only two types of private waste disposal systems are allowed by most code authorities and health departments. They are underground (anaerobic) and above ground (aerobic) systems. Raw sewage seepage, as in cesspools, is no longer permitted, for obvious reasons.

An anaerobic system consists of a closed septic tank and a gray water leach field. When well maintained, this system will last the life of a home. When not properly maintained, a system can fail in five years. Once a system fails, it cannot be reclaimed. A new system will have to be installed at great cost.

Septic tanks

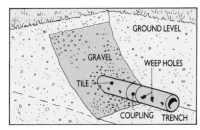

Typical septic tank system
1. Cleanout
2. Ground level
3. Inlet
4. Bacterial crust
5. Outlet
6. Sludge

To understand the need for routine maintenance, you must first understand how a septic system works. As raw sewage is drained into the septic tank, bacteria break the sewage down into gray water, bottom sludge and surface scum. As more sewage enters the tank, gray water rises through a baffle and floats out into the leach field. Once in the field, around 60 percent of the gray water is consumed by plants. The remaining 40 percent is lost through surface evaporation. The nitrate residue left behind is then consumed by another kind of bacteria found only in the top 4 feet of soil.

Because of the scum and sludge left in the tank, your septic tank must be pumped out every two or three years. If you do not have your tank pumped, the sludge at the bottom will rise, thus reducing the capacity of the tank. The scum will continue to build up at the top of the tank until it is deep enough to make its way through the baffle and into the leach field. Once inside the leach field, it will coat the walls of the trench and clog the gravel storage area. When the walls of the leach field are sealed, the leach field will fail.

To keep your septic tank and leach field in working order, have the tank pumped at least once every three years and avoid planting trees near the tank or on top of the leach field.

Repairing a collapsed leach field

Leach field water is transported through perforated pipe (**1**) or under concrete half culverts inverted on concrete blocks (**2**). While perforated pipe is packed in gravel and will not collapse when you drive on it, half culverts often do collapse with the weight of a vehicle. As a general rule, you should not drive over leach fields. If a section does collapse, you can dig it up and replace a broken section with a little shovel work.

Because leach fields work best near the surface, you will find the top of the half culvert only a foot or two down. Dig the dirt above the culvert away and keep it to one side of the ditch. When you hit gravel, dig it out of the trench and store it on the other side.

Each culvert will be 3 to 4 feet long. When you've removed the gravel from around the broken culvert, pull it out. Clean most of the gravel out of the ditch and set a new length of half culvert on the blocks. Shovel the gravel back into the ditch until the dome of the culvert is covered. Then lay several thicknesses of newspaper over the gravel and fill the remainder of the ditch with dirt. Because uncompacted dirt will settle in time, leave a 4-inch mound over the trench. Then soak the dirt and replace the sod.

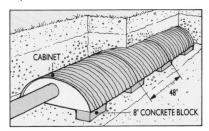

1 Gray water flows in and out of weep holes

2 Inverted half-moon culverts

LAGOON DISPOSAL SYSTEMS

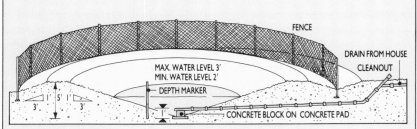

Lagoon disposal system

Aerobic systems are called lagoons. They rely on evaporation and aerobic bacteria to dispose of sewage. The secret to lagoon sewage disposal is in the shape and design of the lagoon. A lagoon is a 3-foot-deep hole with sloping sides and a flat bottom about 30 feet in diameter. The sewer line steps down just before it enters the lagoon and discharges about 3 feet below the water level. Bacteria breaks the solids down and evaporation keeps the lagoon from overflowing. Naturally, lagoons are better suited to windy climates. And for safety reasons, every lagoon must be fenced.

To keep a lagoon working smoothly, keep tall weeds away from the berm so that wind can churn the water and sweep as much moisture away as possible. For the same reason, avoid planting trees in the path of prevailing winds. A lagoon is a self-regulating system that works well with little odor.

SPRINKLER SYSTEMS

With the introduction of polybutylene pipe and fittings to the plumbing market, simple do-it-yourself sprinkler systems are a lot easier to install, even in cold weather climates. Polybutylene pipe will take a freeze, so drainage is not as critical as with other piping materials. Still, the most difficult aspect of any underground sprinkler job is the layout. Plan your work and work your plan.

Installing a system

You will have to make a scale drawing of your lawn and include features such as driveways and sidewalks, which may present piping barriers. You should be able to cover your lawn evenly by matching sprinkler heads to specific areas. Use single-direction pop-up heads for terraces and flower gardens, 45- and 90-degree heads for corners and along drives and 360-degree heads for open spaces. No matter which combination you use, you will have some overlap. Overlapping patterns are not a real problem because coverage is lighter farther away from the heads. By researching the products on the market, you can get a good idea of how each head works and in which situation each should be used. You can do the layout yourself, or you may be able to get help from your local dealer.

Another important factor in planning your system is the water pressure in your home. With high pressure, you may be able to feed all heads from a single line and valve. With less pressure, you may need to divide the system up into two or three separately controlled lines so that one section can be charged at a time. Most sprinkler dealers will lend you a pressure gauge designed to be used on outside faucets. When testing for water pressure, make sure that none of the indoor fixtures is running at the same time.

Burying the lines

To bury the lines, start by marking each head location with a stake. Stretch a string from one stake to another and cut the sod along the string line. Make two cuts in rows about 6 inches apart and about 3 inches deep. Lift the sod up and lay it along the ditch. Then dig an additional 4 to 6 inches deep and level the bottom of the ditch as best you can.

With the ditches ready, roll the plastic pipe along the ditch and cut it at each head and at each intersecting fitting. Connect the tees, elbows and riser pipes at each head location. Then cover the pipe with loose dirt the length of each run. Hold each head level with the ground alongside the riser pipe to determine where to cut the riser. Cut the risers and install each head as you go. All joints will go together with O-ring compression fittings or push-fit fittings.

MAKING THE WATER CONNECTION

With all heads installed, replace the sod all the way up to the house connection. At the house, you will have a decision to make. You can connect directly to the sillcock, or bore through the rim joist or basement wall to connect the pipes inside. A sillcock connection is simpler, but a basement connection gives a cleaner appearance.

To connect to a sillcock, first install a hose-thread-adapted vacuum breaker. A vacuum breaker is necessary to keep your in-house plumbing from back-siphoning contaminated water. It will also allow your heads to drain properly. Then thread the sprinkler adapter to the vacuum breaker and turn on the water. If you need to use the sillcock separately, disconnect the sprinkler fitting temporarily. Of course, in colder climates, the sprinkler system should be disconnected in winter.

Permanent indoor connections

If more than one line must be connected, you are better off connecting inside the basement. Before each line goes into the house, install a backflow preventer. You can use the right angle of the backflow preventer as your elbow into the house. Once inside, run all lines to the nearest ¾-inch water pipe. About a foot before the supply line, install a valve on each line. Then tie the lines together and install a drain valve and another gate valve before tapping into the supply line. This will allow you to drain all lines each winter without shutting off the house water.

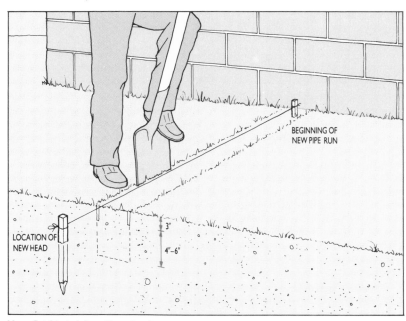

Use a flat-bladed shovel to cut sod

BEGINNING OF
NEW PIPE RUN

LOCATION OF
NEW HEAD

3"

4"–6"

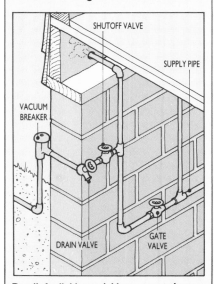

Details for linking sprinkler system to house

SHUTOFF VALVE

SUPPLY PIPE

VACUUM
BREAKER

DRAIN VALVE

GATE
VALVE

PLUMBING OUTDOORS

Outdoor plumbing can be a welcome alternative to stringing garden hoses across your yard. A freezeless spigot in your garden is convenient and a lot less messy than hoses. A seepage pit to serve a drain in your garage workshop can also make life easier. In the final analysis, outdoor plumbing is not much more difficult to install than indoor plumbing. The obvious difference is that in some locations, outdoor plumbing must be able to withstand subzero temperatures.

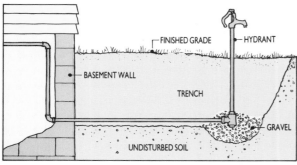

1 A freezeless yard hydrant provides water to remote locations year round

Freezeless yard hydrants

Yard hydrants (1) have their shutoff locations buried below frost level. When you lift the handle at the top of the hydrant, you pull a long stem inside the casing upward. The stopper at the lower end of the stem is lifted out of its seat and water travels up through the pipe casing to the spout. When you push the handle down, the stem pushes the stopper back into its seat and interrupts the flow of water. The water left standing in the pipe casing then drains back through an opening at the bottom of the hydrant, just below frost level.

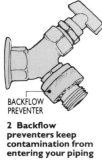

2 Backflow preventers keep contamination from entering your piping

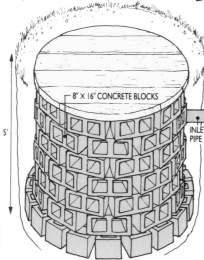

3 Seepage pits dispose of gray water only

Installing a hydrant

Whether you bring water from a basement wall or a buried rural line, you will have to rent a trenching machine. The depth of the trench you dig will be dictated by how deep the ground freezes in your area. Hydrants can be purchased in several lengths for a variety of conditions. A 5-foot depth is common in colder climates.

If you intend to bring water from a house system through a basement wall, simply start the trencher a few inches away from the wall and trench to the hydrant location. At the hydrant location, force the trencher to dig a foot or so deeper. Then, trench a few feet past where the hydrant will be located to avoid having loose soil fall back into the trench.

With the trench ready, measure carefully and drill directly into the open trench from inside the basement. Then slide one end of a coil of buriable plastic pipe through the wall from the outside and reel the coil out in the trench to the hydrant location. Leave the hydrant end of the coil out of the ditch so that you can attach the conversion fitting. The conversion fitting should be a male thread by male insert fitting, preferably made of brass. First, thread the fitting into the hydrant with pipe joint compound. Then, slide two stainless-steel hose clamps onto the plastic pipe and push the insert side of the fitting completely into the pipe. Finally, tighten the clamps around the pipe with a nut driver or small wrench.

Before setting the hydrant down into the trench, pour about 50 pounds of gravel into the section of the ditch that you've made deeper. This will provide a small reservoir for the drain water when the hydrant is in use. Then set the hydrant in place and pour a few more pounds of gravel over the bottom of the hydrant. Before backfilling the trench, seal the wall opening with tar.

Finally, use another insert fitting inside the wall and make the water connection with copper pipe. Because plastic pipe cannot withstand heat, you may wish to use galvanized fittings to get a safe soldering distance away.

If you are splicing into an existing underground plastic line, installation is much the same. The only real difference is that instead of using a copper wye insert fitting, you will use an insert tee. The piping size you use will depend upon the hydrant's inlet size and your system's water pressure. A 1¼-inch line is commonly used.

Preventing backflow

Because both sillcocks and yard hydrants are frequently used to apply lawn and garden chemicals, water contamination becomes an issue. Contaminated hoses can easily backwash when fire hydrants are opened or when other high-volume draws are made on city piping near your home. This is especially true when a section of city main is shut down for repairs. To protect yourself and your neighbors, you should install backflow preventers (2) on all outdoor faucets. Backflow preventers allow water to pass in one direction only, thereby checking any system backwash.

Installing seepage pits

Seepage pits (3) are miniature leach fields designed to dispose of gray water discharged from appliances and floor drains. If your drainage system is served by a municipal sewer, seepage pits will not be needed, and in most cases, not allowed. If, however, yours is a home built in the country, a seepage pit can save you the trouble of tapping into your septic system when outbuildings are some distance from the house.

The size of the pit you build will depend upon anticipated volume. For a garage or workshop, a 3-by-4-foot inside diameter is often sufficient. Before starting the job, however, check with local code authorities for structural guidelines.

Start by digging a more or less round pit roughly 5 feet deep and 5 feet in diameter. Then lay a starter course of concrete blocks side-by-side around the outside walls of the pit. Follow with a series of courses until you are within 1½ feet of grade. At this point, you can trench the drainage pipe to the pit. If you slide the drainpipe through an opening in one of the top blocks, it will be held permanently in place. Finally, construct a cover from treated lumber.

For sandy soil, you may wish to tape layers of newspaper around the outside perimeter of the block to keep the soil from sifting in when you backfill. The newspaper will decompose after the soil settles.

A well-functioning fireplace is the product of many design elements working together in harmony. When fireplace smoke drifts outward into the room rather than escaping smoothly up the chimney, there may be several causes. If the fireplace has *always been a smoky one, chances are that its construction includes one or more design flaws. If the condition is a recent development or is intermittent, solving the problem may mean only simple adjustment or cleaning.*

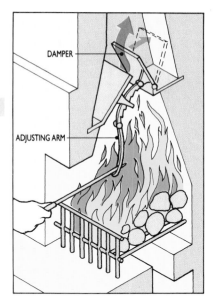

Regulating updraft
Regulate chimney updraft by adjusting damper. Operate adjusting arm using poker when fire is burning.

SEE ALSO

Details for: ▷
Sweeping chimneys   54

Remedying simple problems

If the fireplace smokes only occasionally, or has just begun to smoke, run through this checklist of minor adjustments. First, be certain that when the fire is burning, the damper is fully open. Most dampers can be adjusted when hot by pushing the protruding end of the handle with a poker. Check, too, that the chimney is free of obstructions, especially if a normally clean-burning fireplace suddenly starts smoking. Along with this, make sure that the chimney is kept regularly swept. Accumulated soot can eventually cause smoking and, worse, a chimney fire.

Check that the fire is built well back in the firebox so that no burning logs project beyond the fireplace opening. Try raising the height of the fire several inches by placing the logs on a grate elevated on firebricks. Inadequate inflow of air—too little to support combustion or feed the chimney's updraft—may also be the culprit, especially if you have altered the ventilation pattern in the room by adding insulation and weatherstripping. To alleviate this cause of a smoky fireplace, open a door or window to admit more air.

Correcting chimney faults

To produce an updraft, air must flow steadily across the opening at the top of a chimney, creating a partial vacuum within, which aids in drawing the heated air from the fireplace upward. In order for the air to be unobstructed as it flows, a chimney must be at least 3 feet higher than any object within a 10-foot radius, including roof peaks, trees, television antennas or other chimneys. If you cannot increase the height of a

too-short chimney by adding to it, attaching a chimney cap or smoke puller (a fan mounted in the chimney opening) may help. Before undertaking such modifications, consult a professional chimney mason.

Uncapped chimneys should have a sloping cowl of mortar on all four sides to direct passing air upward and over the opening. Otherwise, air striking the chimney will eddy and swirl erratically.

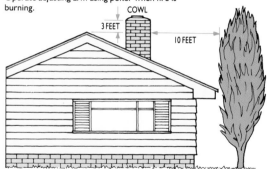

Minimum safe clearance for chimneys

Correcting fireplace proportions

To draw smoke upward properly, the dimensions of the chimney flue must bear a certain proportional relationship to those of the brick-lined or steel-lined area where the fire is actually built, called the firebox. Also, the firebox itself must be built to a certain shape, in order to both reflect heat outward yet direct smoke upward. Often, either the firebox is built too large for the flue rising inside the chimney, or the flue (usually a retrofit inserted into a chimney that was originally built without one) is too small.

One solution is to fit a metal fireplace hood, available from fireplace and woodstove supply stores and some home centers, across the top of the fireplace opening to decrease its overall size and also trap smoke that might otherwise seep out. To install the hood, first determine how large it must be by

holding a piece of metal or dampened plywood across the top of the fireplace when a fire is burning. As the fireplace begins to smoke, gradually lower the sheet of material until the smoking is contained. Purchase a hood this size. Most hoods attach to the fireplace surround by means of special masonry hangers.

Another solution is to fit glass doors across the entire front opening of the fireplace. These, too, are available in many sizes from fireplace and woodstove stores, as well as home centers. Although some heat may be lost to the room when the doors are closed the fire is entirely enclosed and smoke completely contained. (There are, however, energy-efficient models which actually enhance the amount of heat produced by the fireplace.)

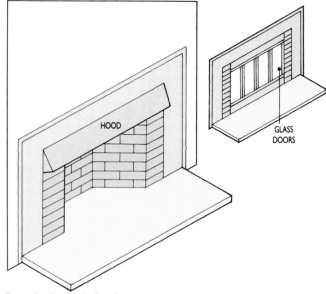

Remedies for smoky fireplaces
A hood or glass doors can prevent smoke from entering a room.

REMOVING AN ANTIQUE FIREPLACE

To take out an antique fire surround and hearth is easy enough, but it can create much dust and debris. Any hammering is likely to cause quantities of soot to fall down the chimney. Before you start, sweep the chimney, move all furniture as far from the fireplace as possible, roll back the carpet and cover everything with drop-cloths. There is a good demand for Victorian fire surrounds and some are valuable. If you remove yours undamaged, you may be able to sell it.

Removing the hearth

Most old-fashioned hearths were laid after the fire surround had been fitted and so must come out first, but check beforehand that your surround has not been installed on top of the hearth.

Wear safety goggles and heavy gloves against flying debris and use a 2-pound sledgehammer and bricklayer's chisel to break the mortar bond between the hearth and the subhearth below.

Knocking in wood wedges will help. Lever the hearth free with a crowbar or the blade of a strong garden spade and lift it clear. It will be heavy, so get someone to help.

Some older hearths are laid level with the surrounding floorboards and have a layer of tiles on top of them. Here all that's needed is to lift the tiles off carefully with a bricklayer's chisel.

Removing the surround

Most fire surrounds are held to the wall by screws driven through metal lugs set around their edges. They will be concealed in the plaster on the chimney breast. To find the lugs, chip away a 1-inch strip of plaster around the surround, then expose the lugs completely and take out the screws. If they are rusted and immovable, soak them in penetrating oil, leave for a few hours and try again. If that fails, drill out their heads. The surround will be heavy, so have some help available when you lever it from the wall and lower it carefully onto the floor (see left).

Brick and stone surrounds
A brick or stone surround can be removed a piece at a time, using a bricklayer's chisel to break the mortar joints. There may also be metal ties holding it to the wall.

A wooden surround
A wooden surround will probably be held by screws driven through its sides and top into strips fixed to the chimney breast inside the surround. The screw heads will be hidden by wooden plugs or filler. Chisel these out, remove the screws and lift away the surround.

• **Saving a fireplace**
Fire surrounds can be very heavy, especially stone or marble ones. If you wish to keep yours intact for sale, lay an old mattress in front of it before you pull it from the wall so it will be less likely to break if it should fall.

Taking out a fireplace
1 The hearth chipped free
2 The subhearth at floor level
3 The fireback, to be broken out
4 The fire surround; a brick or stone one can be taken out in bits
5 Metal lugs hold most surrounds in place

REPLACING CRACKED TILES

Cracked or broken tiles in a hearth or fire surround should be replaced with sound ones, but you may not be able to match those in an old fireplace. One solution here is to buy some new tiles that pleasantly complement or contrast with the originals and replace more than just the damaged one or two, making a random or symmetrical pattern.

Break out the damaged tile with a hammer and cold chisel, working from the center outwards. Wear thick gloves and safety goggles against flying bits of tile and protect nearby surfaces with dropcloths. When the tile is out, remove all traces of old adhesive or mortar and vacuum up the dust.

If necessary, cut the new tile to shape. Spread heat-resistant tile adhesive thickly on its back and on the surface where it is to go. Don't get adhesive on its edges or the edges of surrounding tiles. Set the tile in place, taking care that the clearance is equal all around, and wipe off any excess adhesive. Leave it to set and then apply grout.

If you are replacing only one tile, it's not worth buying a tub or packet of adhesive. Instead, mix a paste from coarse sawdust and wood glue, which will work just as well. If the tile is very close to the fire you can use some fire cement.

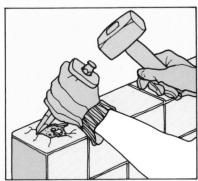

Chipping out a damaged tile
Start in the middle and work out to edges. Clean out all old mortar or adhesive.

Complete retiling

If a lot of the tiles are damaged or crazed, your best course may be to retile the surround and hearth entirely. This is much less trouble than it sounds, as you can simply stick the new tiles directly on top of the old ones. First make sure that the old tiles are clean and remove any loose pieces, then apply your tile adhesive and stick the tiles on in the ordinary way.

INSTALLING AN OLD-FASHIONED FIREPLACE

In the past 30 years or so traditional fireplaces have vanished from many older houses, swept away in the name of modernization. But now they are being appreciated again and even sought after. You can reinstate an old-fashioned fireplace as described here.

Most period fire surrounds are held in place by lugs screwed to the wall, but some can be attached with mortar. A plaster surround can be held with dabs of bonding plaster.

First, remove a strip of plaster from around the fire opening about 2 inches wider all around than the surround.

If the surround incorporates a cast-iron centerpiece, it must be fitted first. Most of them simply stand on the back hearth, but some have lugs for screwing to the wall. If yours has lugs, use metal wall plugs or expanding bolts. Fit lengths of asbestos-substitute packing as expansion joints where the grate or centerpiece touches the fireback.

Hold the surround in place, mark the wall for the screw holes and drill them. Use a level to check that the surround is upright and the mantel horizontal, and make any needed corrections by fitting wooden wedges behind the surround or bending the lugs backward or forward.

An alternative method for plaster surrounds is to apply mortar or plaster to the wall and prop the surround against it with boards until the mortar or plaster sets.

Replace the hearth or build a new one. Set the hearth on dabs of mortar and point around the edges with the same material.

Replaster the wall and fit new baseboard molding between the hearth and the corners of the chimney breast.

Hold a light surround in place with boards while the plaster sets

SEALING A FIREPLACE OPENING

If you have removed an old fireplace, you can close the opening by bricking it up or by covering it with plasterboard. The latter will make it easier to reinstate the fireplace if you want to at some time in the future. In either case, you must fit a ventilator in the center of the opening just above baseboard level. This will provide an airflow through the chimney and prevent condensation from forming and seeping through the old brickwork to damage wall decorations.

Restoring the floor

If the floor is solid, you need only bring the subhearth up level with it, using cement. You can also do this with a wood floor if it is to be carpeted. If you want exposed floorboards, the subhearth will have to be broken away with hammer and cold chisel to make room for a new joist and floorboards to be fitted.

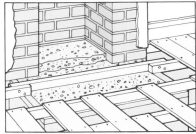

A new joist for extended floorboards

Sealing the opening with bricks

If you wish to brick up the firebox opening, remove bricks from alternate courses at the edges of the opening so that the new brickwork can be "toothed in." Provide ventilation for the chimney by fitting a brick vent centrally in the brickwork and just above baseboard level. Plaster the brickwork and allow it to dry out thoroughly before you redecorate the wall.

Finally, lever the old pieces of baseboard from the ends of the chimney breast and replace them with a full-length piece from corner to corner.

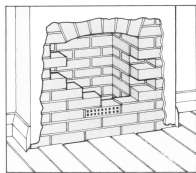

Install a brick vent at baseboard level

Sealing the opening with plasterboard

Cut a panel from ⅜-inch plasterboard and nail it onto a wood frame mounted inside the fire opening.

Use 2 × 2-inch sawn lumber for the frame. Nail it into the opening with masonry nails, setting it in so that when the plasterboard is nailed on, it will lie flush with the surrounding wall if it is to be papered. Place it ⅛ inch deeper if a plaster skim is to be added. For papering, fix the plasterboard with its ivory side out; for a plaster skim, the gray side should be showing. After decorating or plastering the panel, fit a ventilator.

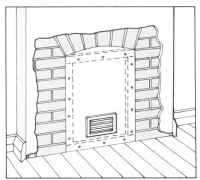

Attach plasterboard panel to inset wood frame

Closing off the chimney top

When you close off a fireplace opening, you will have to cap the chimney in such a way as to keep rain out while allowing the air from the vent in the room to escape. Use a half-round ridge tile bedded in cement, or a metal cowl, either of which will do the job.

Half-round ridge tile

Commercial cowl

INSTALLING A COMBINATION INSERT

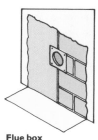

Flue box
Connecting the flue of a log-burning stove to a flue box set into a bricked-up fireplace may be preferred to using a horizontal or vertical closure plate. A flue box is a cast-iron frame with a hole in the center for the flue outlet.

Easy-to-install fireplace inserts, consisting of an energy-efficient tubular fireplace grate and blower unit plus glass doors that fully cover the fireplace opening, are widely available and can significantly boost the amount of heat produced by an ordinary fireplace.

Although there are many different varieties, installation of most combination inserts involves first assembling the grate, usually by bolting the convection tubes in sequence to the grate supports which elevate the unit off the firebox floor, and then fitting the unit into the firebox, to be followed by the assembled doors. Many grates are free-standing, and may be merely slid in and out for periodic cleaning.

The tops of the convection tubes generally seat against a vented portion along the top of the door frame, so that the heated air rising through them as the fire burns is directed outward. A blower assembly attaches to the lower portion of the door to draw air needed for combustion into the lower ends of the tubes and assist in forcing it upward and out again, into the room.

The doors themselves fit around the perimeter of the fireplace surround and are held in place with masonry bolts. Installation is usually a matter of drilling into the masonry, installing bolt inserts, then attaching the door frame using the bolts supplied with the unit. In addition, most manufacturers recommend sealing the gap between the insert frame and fireplace surround with fiberglass insulation to prevent heat leaks.

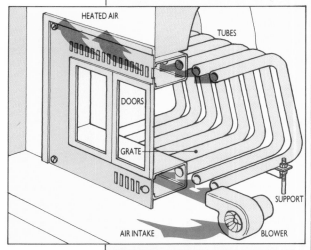

Combination tube and glass-door insert
This installs easily and increases fireplace efficiency.

WOOD HEATERS AND INSERTS

A modern enclosed fireplace, or room heater, can be freestanding or inset (built-in). Both are very efficient at heating individual rooms and, with the addition of back boilers, can provide domestic hot water and central heating, too. The toughened-glass doors of closed fireplaces and inserts, which allow the glow of the fire to be seen, open for extra fuel to be added.

A freestanding heater on the hearth

Freestanding room heaters are designed to stand on the hearth forward of the chimney breast. They radiate extra warmth from their casings but their size can make them obtrusive in small rooms. You may also have to extend the hearth to the required 18 inches in front of the heater.

A heater of this type has a flue outlet at its rear which must be connected to the chimney, and the easiest way of arranging this is to seal the outlet into a metal backplate that closes off the fire opening. The projecting end of the outlet must be at least 4 inches from the back wall of the firebox.

The closure plate should be of steelboard (metal-covered asbestos), which is sold at woodstove supply stores. Use metal wall plugs to hold the screws, and seal the joint between plate and opening with asbestos-substitute packing and fire cement. Alternatively, stovepipes can be fitted into the fireplace damper opening after removing the damper plate.

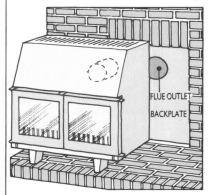

A backplate closes off the fire opening

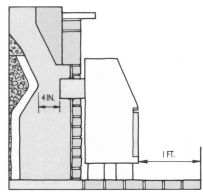

Key measurements for an enclosed fireplace

A freestanding heater in the fireplace

Some freestanding room heaters are designed to stand in the fire opening. This type of heater has a flue outlet in its top which must be connected to a closure plate set in the base of the chimney.

The plate can be of metal or precast concrete. For access to fit it, remove some bricks from the chimney breast just above the opening but below the load-bearing lintel. If the plate is of concrete, take out a course of bricks around the bottom of the chimney to support it properly. You can insert a metal plate into a chased-out mortar joint or fasten it with expansion bolts. Bed the plate on fire cement, sealing the edges above and below. Check that the heater's outlet enters the chimney flue, and seal the plate joint with asbestos-substitute packing and fire cement.

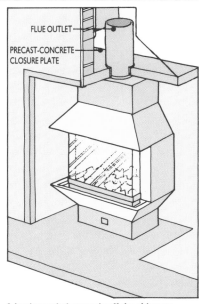

A horizontal plate seals off the chimney

An inset room heater

An inset room heater has its flue outlet mounted on top to be connected to a chimney closure plate or stovepipe rising through the chimney.

This type of appliance is designed to fill and seal the fire opening completely, so to install one you may have to modify your present fire surround or, if the opening is very large, even build a new one. The sides of the surround must be exactly at right angles to the hearth, as the front portion of the heater's casing has to be sealed to both. If the surround and the hearth form an odd angle, a good seal with the heater casing will be impossible. The seal is made with asbestos-substitute packing material.

More inset room heaters are screwed down to the firebox floor, and some may need a vermiculite-based infill around the back of the casing which must be in place before the chimney closure plate is fitted and the flue outlet connected.

Some come supplied with their own fire surrounds, complete with drop-in closure plates designed to make their installation easier.

Finish the job by restoring the brickwork of the chimney breast and replastering it if necessary.

One of the most economical ways to keep a room warm is by means of a modern slow-combustion log-burning stove—if you have access to cheap wood. Like fireplace inserts or inset room heaters,

they can be stood on the hearth with rear flue outlets or installed in the fireplace with top-mounted outlets. A good log-burning stove can burn all day or night on one load of wood.

A log-burning stove is best installed forward of the chimney breast so that you get the full benefit of the heat that radiates from its casing. You can stand it on your present hearth provided that the hearth projects the required minimum of 18 inches in front of the stove and at least 12 inches on each side of it (36 inches from the nearest combustible surface). Otherwise you will have to make a new and bigger hearth. The hearth must be level and constructed of stone, brick or tiles.

A log-burning stove can be fitted with a stainless-steel, insulated flue pipe and this can be passed through a vertical back closure plate that seals off the whole fireplace opening or through a horizontal plate that closes off the base of the chimney. It may even be passed

through a fitting—called a thimble— in the face of the chimney itself.

As an alternative to all this you can brick up the fireplace completely and install a flue box (see opposite).

If the flue pipe is connected to a stovepipe, the connection should include an extension of pipe below the joint, to serve as an ash pit. The installation must incorporate good access for efficient chimney sweeping, which will be needed often.

It's important to follow local fire codes and building regulations when installing a wood stove. These generally include setting the unit at a safe distance from combustibles, making a positive connection between the stovepipe and flue and being sure the flue extends at least 36 inches above the roof.

SEE ALSO

Details for: ▷	
Sealing fireplace	49
Closure plate	50

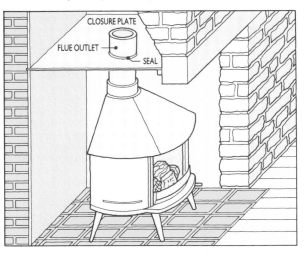

Vertical flue outlet
This type of stove has its flue sealed into the opening of a horizontal closure plate in the base of the chimney.

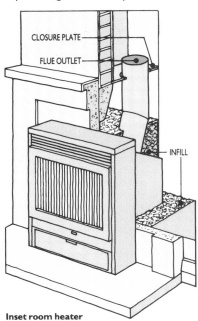

Inset room heater
The top-mounted flue outlet connects to a horizontal closure plate in the chimney base. Some versions need an infill around the rear casing.

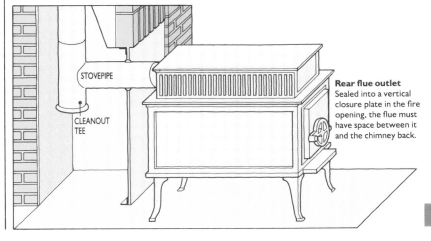

Rear flue outlet
Sealed into a vertical closure plate in the fire opening, the flue must have space between it and the chimney back.

INSTALLING A FLUE LINER

If your house was built before World War II, there's a good chance that its chimney is unlined and is simply a rectangular duct whose brickwork is either stuccoed with cement or quite exposed.

Over the years corrosive elements in the rising combustion gases eat into the chimney's mortar and brickwork and weaken it, allowing condensation to pass through and form damp patches on the chimney breast and, in extreme cases, letting smoke seep through. This is particularly true where coal or wood-burning appliances are in use.

Choosing a flue liner

You can deal with these problems by installing a flue liner, which will prevent the corrosive elements from reaching the brickwork. It will also reduce the "bore" of the flue, and that will speed up the flow of gases and prevent their cooling and condensing. The draft of air through the flue will improve and the fire will burn more efficiently.

However, it is important to fit the type of liner that's appropriate to the kind of heating appliance being used and to be sure the size is large enough to prevent the fireplace from smoking. Ask the appliance supplier or building inspector. Linings are tubes, one-piece or in sections, of metal or other rigid, noncombustible material.

ROOF ACCESS

You can rent easy-to-use light alloy roof scaffolding. Two units will make a half platform for a central or side chimney; four will provide an all-around platform.

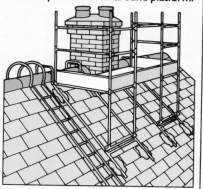

Scaffolding is essential for safe working

Installing a one-piece flue liner

A popular type of liner is a one-piece flexible corrugated tube of stainless steel that is easily fed into a chimney that has bends in it. Unfortunately, this type of liner is not suitable for use with coal or wood-burning appliances. To install it you must get onto the roof and erect scaffolding around the chimney (see above right).

First sweep the chimney, then chop away the mortar around the base of the chimney pot with a hammer and cold chisel, remove the pot—it will be very heavy, so take care—and lower it to the ground on a rope. Clean up the top of the chimney to expose the brickwork.

The liner is fed into the chimney from the top. Drop a strong weighted line down the chimney (**1**) and attach its other end to the conical endpiece of the flue liner. Have an assistant pull gently on the line from below while you feed the liner down into the chimney (**2**). When the conical endpiece emerges below, remove it and connect the liner to a closure plate set across the base of the chimney or to the flue outlet of the heating appliance, and seal the joint with an asbestos-substitute packing and some fire cement.

Return to the roof, fit the top closure plate and bed it in mortar laid on the top of the chimney, adding extra mortar to match the original (**3**). Finally, fit a cowl to the top of the liner, having chosen one appropriate to the heating appliance being used (see left).

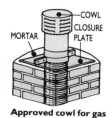

Approved cowl for gas

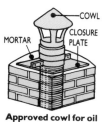

Approved cowl for oil

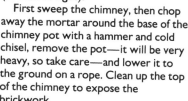

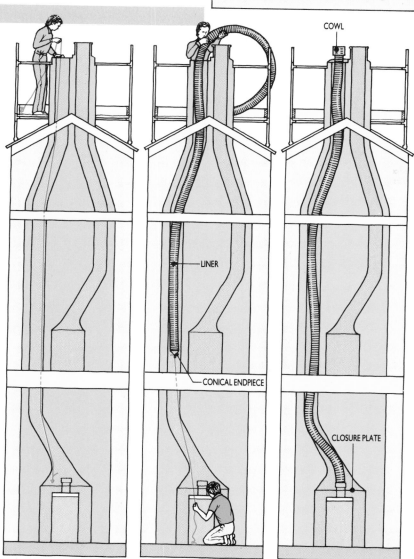

I Lower attached line **2 Feed liner to helper** **3 Complete top closure**

Installing a sectional flue liner

A sectional flue liner has the space around it filled with a lightweight concrete that strengthens and insulates the chimney but needs good foundations for the added weight. Like the one-piece liner (see opposite), it is inserted from the top. First, cement a steel base plate across the bottom of the chimney.

Tie the first flue section to a rope and lower it into the chimney. Connect the next section to it by one of the steel collars supplied and lower the two farther down. Continue adding sections and lowering the liner until it reaches the base plate, then seal it in place.

If there are any bends in the chimney, you will have to break into it at those points to feed the sections in. This may be a job for a professional.

For filling in the chimney around the liner use concrete made with a lightweight aggregate such as expanded clay or vermiculite. Pour this into the chimney around the liner and finish off the mortar cap at the top.

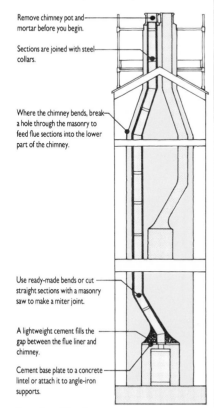

Remove chimney pot and mortar before you begin.

Sections are joined with steel collars.

Where the chimney bends, break a hole through the masonry to feed flue sections into the lower part of the chimney.

Use ready-made bends or cut straight sections with a masonry saw to make a miter joint.

A lightweight cement fills the gap between the flue liner and chimney.

Cement base plate to a concrete lintel or attach it to angle-iron supports.

A sectional flue liner
Installing a sectional flue liner can be such a complicated procedure that it is worth asking for professional quotes before you decide to tackle the job yourself.

Though fireplace surrounds and hearths get a great deal of wear and tear, they will keep their appearance if they're looked after and if certain elementary care is taken. For example, avoid standing cups, glasses and ashtrays on them and be especially careful with such drinks as coffee, tea, alcohol and fruit juices, which can be very damaging to some surfaces, particularly marble.

Always let the hearth and surround cool down before you clean them, and avoid standing on the hearth when you clean the surround.

The different materials respond to different cleaning methods.

STONE

This can be sponged off with warm water that has a little detergent mixed into it. You can remove the more stubborn stains with a stiff brush.

MARBLE AND GRANITE

It is easy to damage the fine finish on these stones, so they should be treated with some care. Wash them regularly with warm, soapy water and polish them with a chamois leather or a good-quality wax polish. Patch small chips with a putty made from kaolin powder (china clay) and epoxy glue. Rub down the hardened filler with silicon carbide paper and touch up with lacquer.

CERAMIC TILES

Wash these with warm water containing a little detergent. Never use any kind of abrasive cleaner on them.

BRICKS

Dust bricks off occasionally with a soft brush, but never use soapy water on them. If you have built your own brick surround, you should treat it with a commercial sealer to prevent "dusting." Broken bricks can be stuck together again with an epoxy adhesive, or you can cut them out and replace them with new ones, using fresh mortar.

SLATE

Wash this with warm water mixed with a little detergent, using a stiff brush. If the slate is unpolished, remove stubborn stains with an abrasive cleaner.

METALWORK

Wash this with warm, soapy water and take off stubborn tar and soot stains with commercial metal cleaner. Clean up cast-iron surrounds with wet-and-dry abrasive paper or emery cloth, then finish with stove blacking.

WOODWORK

If the grain is exposed, maintain the finish with a good wax polish. Fill any cracks or gouges with a commercial wood filler tinted to match with a little wood stain. The woodwork can also be painted in the ordinary way.

Stone surround

Marble surround

Wood surround

SEE ALSO

Details for: ▷	
Chimney base plate	50

● **Casting a flue liner**
Professional installers can cast a flue liner in place. A deflated tube is lowered into the chimney. It is inflated, and a lightweight infill is poured into the gap between the tube and chimney. When the infill has set, the tube is deflated and removed, leaving a smooth-bore flue.

53

OPEN FIRES

For centuries open fires were our only domestic heating. Inefficient and wasteful, their only benefits were the radiant heat from the burning fuel and some milder *warmth from heated chimneys. They are nowadays used mainly as attractive focal points in homes heated by more modern means.*

How an open fire works

To burn well, any fire needs a good supply of oxygen (1) and a means for its smoke and gases to escape (2). If either of these is cut off the fire will be stifled and will eventually go out.

The domestic open fire is built on a barred grate (3) through which ash and debris fall into a removable tray and oxygen is sucked up into the base of the fire to maintain combustion.

As the fuel burns it gives off heated gases which expand and become lighter than the surrounding air so that they rise (4). To prevent the gases and smoke from drifting out and filling the room, a chimney above the fire gives them an escape route, taking them above the roof level of the house to be harmlessly discharged into the atmosphere beyond.

As the hot gases rise they cause the suction at the bottom of the fire which draws in the supply of oxygen that keeps it burning. For this reason, a good fire needs not only an effective chimney but also good ventilation in the room where it is burning so that the air consumed by the fire can be continually replenished. Sometimes efficient draftproofing at doors and windows can cause problems with a fire, and prevent its burning properly by denying it the constant supply of air that it needs. In such a case, the ventilation must be provided by means of a vent, or opening a window slightly; underfloor ventilation is another possibility.

The simple workings of a traditional open fire
1 Air is sucked in as the gases rise
2 Gases escape up the narrow flue
3 The grate lets ash out and air in
4 Gases vent to the air outside

CHIMNEY FLUE
SMOKE SHELF
DAMPER
GRATE
FIREPLACE
CLEAN-OUT DOOR
ASH PIT

Sweeping chimneys

- **Vacuum sweeping**
You can rent a special vacuum cleaner for chimney sweeping. Its nozzle is inserted from above or below and sucks out the soot—a very clean method but no use for heavy soot deposits or other obstructions.

- **Chemical cleaning**
There are chemicals which remove light soot deposits and stop further sooting up. In liquid or powder form, sprinkled on the hot fire, they make a nontoxic gas which causes soot to crumble away from the chimney sides.

All solid fuels give off dust, ash, acids and tarry substances as they burn, and this material is carried up through the chimney, where a part of it collects as soot. If too much soot collects in a chimney, it effectively reduces the diameter, and therefore the gas flow, and prevents the fire from burning properly. It can even cause a blockage, particularly at a bend, or the more serious hazard of a chimney fire.

To prevent soot build up, sweep your chimneys at least twice a year, once during the heating season and once at the end to prevent acids in the soot from attacking the chimney's lining and mortar joints during the summer. If a chimney is left unswept for too long a period, the consumption may increase, smoke may start billowing into the room and soot may occasionally drop into the fire.

Though it's seemingly a dirty job, you can sweep a chimney without making a great deal of mess to be cleaned up afterwards, provided you take some care. You can rent the brushes. The modern ones have nylon bristles and "canes" made from polypropylene.

Remove all loose items from the fire surround and the hearth, then roll back the carpet and cover it with a dropcloth or newspapers for protection. Drape a large old sheet or blanket over the fire surround, weighting it down along the top and leaning something heavy against each side to form a seal with the edges of the fire surround.

Actual sweeping may be done by pushing a brush up from the fireplace, or by forcing it down from the chimney top. If the roof is dangerous or the chimney is covered by a nonremovable chimney cap, sweep from inside the house (you may have to remove the damper plate at the base of the flue to fit the brushes). If possible, sweep from the top down. To do this, first be sure the fireplace opening is tightly covered with dropcloths. From the roof, insert a correct size brush (it should fit tightly) into the chimney top and push it down toward the fireplace opening below. Screw on additional lengths of cane as necessary to reach the proper depth. Work the brushes up and down, being careful not to damage any mortar inside the chimney. When you reach the bottom, withdraw the brush. Wait one hour for the dust to settle, then vacuum the debris from the fireplace floor and smoke shelf using a heavy-duty industrial vacuum cleaner available from rent-it centers.

Though using a brush and canes is the most time-honored—and the most effective—way of sweeping a chimney, in recent years some other methods have been found for doing this dirty job (see left).

Cleaning from above
Insert brush and canes at chimney top. Brush up and down, being careful not to disturb mortar joints.

Cleaning from below
Seal off the fireplace with an old sheet and feed the canes up under it.

CENTRAL HEATING SYSTEMS

A central heating system supplies heat from a single source to selected rooms—or all the rooms—in the house. It is a much more efficient arrangement than having an individual heater in each room as it has only one appliance to be controlled, cleaned and maintained.

Categories of heating systems

Central heating systems are categorized by the medium used to deliver the heat from its source to the various outlets around the house. The three most common systems are forced-air, circulating hot water, and steam. Heat sources for these are normally a gas- or oil-fired furnace (though in some areas coal- and even wood-fueled systems are still in use), which in turn either heats water in a boiler or air passing directly past the flame. Electric heating systems, which derive heat through simple resistance wiring usually embedded in ceilings or floors, or which feed heated air through a blower, are a less popular system, due to expensive energy costs. Their advantages, however, are cleanliness and high efficiency.

Forced-air systems

Modern forced-air heating systems consist of a furnace which heats air, a large squirrel-cage blower which circulates that air, and a system of air ducts through which the heated air is directed throughout the house. A secondary system of ducts is also part of

the system. Through it, cooled air returns to the furnace.

Because of the size and unwieldy nature of the air ducts, forced hot-air heating systems are almost always installed during new construction. Rarely are they feasible as retrofits. From the main duct leading away from the blower chamber the delivery ducts branch off, running between floor joists and wall studs to their openings at louvered registers normally located in outside walls, a short distance up from the floor. Cool air returns via larger-size ducts, which generally open directly into the floor, most often near inside walls or near the center of a room, but also in natural air traps such as stairwells. Since hot air rises and cold air descends, the warm air rising along the outside walls heats the rooms; then, as it cools, descends and enters the return ducts, creating a convection current which serves the entire enclosed area.

In large systems, the ductwork is usually designed to produce heating zones, groups of rooms or areas served by a single branch-duct system which can be isolated from the rest. By the use of dampers which physically close off key ducts (dampers are operated manually or by thermostats), the amount of heat directed throughout the house can be adjusted so that unused rooms receive less heat, while frequently inhabited rooms receive more.

BALANCING A FORCED-AIR SYSTEM

Adjusting the air flowing through a forced-air heating system—a process called balancing—assures that each room receives the most comfortable amount of heat. Whereas the registers at the ends of each duct run are generally louvered and adjustable, it is best to use these merely as a means of directing air up or down within a room. To actually control the amount of air passing through the registers, most duct systems are fitted with metal plates called dampers inside strategically located sections. Dampers are controlled by means of a handle on the outside of the duct. Turning the handle so it is parallel with the run of the duct allows maximum airflow; turning it toward perpendicular, the less air flows through.

To balance the system, you must merely adjust the dampers until the appropriate settings are achieved. The process is simple but, because each room requires 6 to 8 hours to become properly heated, takes time. Balance the system during a cold spell, when the furnace is running at its peak. Begin by closing down the damper to the most uncomfortably hot room nearest the furnace. This will send correspondingly greater amounts of heat to more distant registers, so, after waiting the required heating-up time, move on to the next uncomfortably hot room and adjust the damper there. You may determine the desired temperature of a given room by "feel," or by holding a thermometer in the room a few feet above floor level.

After you have moved through the entire house and adjusted all the dampers once (a process that may have taken a week or more), go back and fine-tune your work by making minor adjustments to the dampers that seem to require it. Wait the necessary time between each. When you are finished, mark with white paint on the duct the positions of all the damper handles so that they are permanently recorded.

If, after balancing, there exist rooms at the far end of duct runs that receive insufficient heat, a common solution is to increase the speed of the squirrel-cage blower. This is done by adjusting or replacing one of the drive pulleys on the motor. However, since increasing blower speed places additional strain on the motor, consult a furnace repair person beforehand.

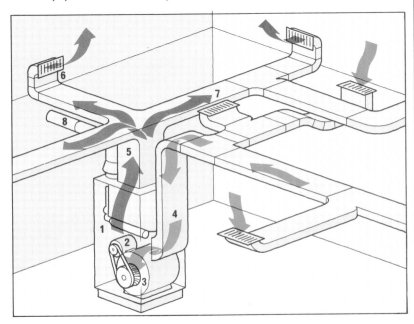

Typical gas-fired forced-air system
1 Furnace (gas)
2 Motor
3 Blower
4 Cold-air return
5 Warm-air delivery
6 Warm-air registers
7 Cold-air registers
8 Exhaust (to chimney)

CENTRAL HEATING SYSTEMS

Circulating hot-water

The most popular form of central heating is a circulating hot-water system. Water is heated to between 120 and 180 degrees F in a furnace-fired boiler and then is forced by a circulator pump through a system of pipes leading to and from radiators or convectors located throughout the house. Some layouts pump water through a single loop of pipe, off which branch piping both feeds hot water and returns cooled water to each radiator in sequence. With these systems, careful balancing (as for forced-air systems) is necessary to assure that radiators at the far end of the pipe loop obtain sufficient heat. A better type of layout is one made up of two sets of pipe runs, one to carry only the hot water, and one to return only the cooled. These systems require far less adjustment, since water flowing to the farthest radiator does not become mixed with cooled water returning from radiators along the way.

Hot-water systems include an expansion tank—normally located near the boiler—which contains air. As water in the system becomes heated and expands, the air in the tank is compressed, in turn placing the water in the system also under pressure, and thus preventing it from becoming steam.

Typical circulating hot-water system
1 Boiler (oil-fired)
2 Expansion tank
3 Circulator pump
4 Hot-water supply
5 Control valves
6 Thermostats
7 Convectors
8 Cooled-water return
9 Main water supply

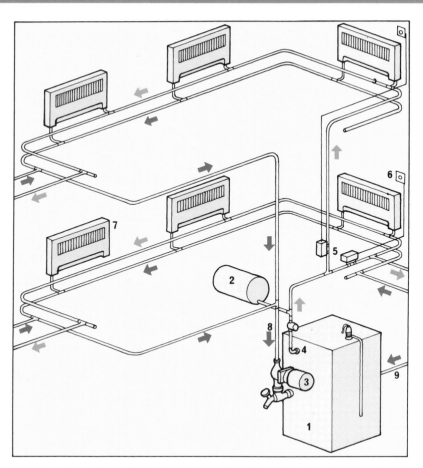

Steam heat

Few contemporary homes are built with steam heating systems. However, they are still to be found in many older homes, especially those built prior to World War II. Steam systems operate much like a single-pipe circulating hot-water system (see above). Water heated in a boiler until it becomes steam travels under its own pressure through a single pipe forming a loop around the house, off which branch plumbing services each radiator. As the steam cools by giving up its heat to the radiators, it changes back to water and returns to the boiler by gravity via the same pipe from which it was originally dispersed. Steam systems require neither a circulator nor an expansion tank. Piping must, however, slope downward from all points toward the boiler, and balancing is necessary to properly distribute steam to each room. In addition, steam radiators must be frequently drained of both air and water to remain in working order.

Typical steam/heat system
1 Boiler (oil-fired)
2 Steam supply/return pipe
3 Radiator
4 Main water supply

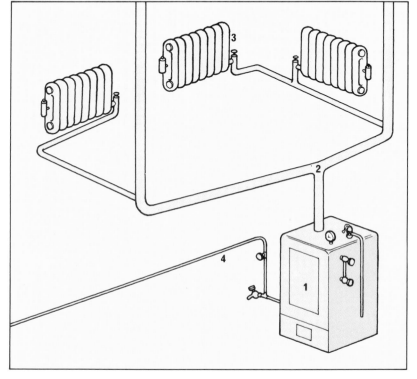

CENTRAL HEATING SYSTEMS

Electric heat

Electric central heating systems have long been popular in Europe and have been available in the U.S. for approximately fifty years. Their popularity reached its peak during the 1960s, prior to the worldwide increase in energy costs, which has subsequently caused large-scale electric heating systems to be expensive to operate. Still, electric heat offers distinct advantages of comfort—quick, efficient, draft-free warmth—and, since no fuel is actually burned, the lack of need for such apparatus as a chimney, fuel storage

tank, and in many systems even ductwork, piping, and vents.

Because most electric systems operate on the principles of radiant, rather than convective, heat, they provide the most uniform heat and achieve their greatest energy efficiency when installed over the greatest possible area. Installing radiant heating is similar to installing electric lighting: Many small, low-intensity units covering a broad area produce better results than only a few high-intensity units widely spaced.

Embedded tubes and cables

In new-floor and -ceiling construction, grids of electric resistance heating cable are often embedded between layers of whatever building material is used, with insulation placed below (in the case of floors) or above the wiring (in the case of ceilings) to direct all heat toward the living area. Each grid forms a separate heating zone and is wired as an individual circuit to a central service panel (generally not the same panel that controls the other household circuitry), and is controlled by an individual thermostat.

In place of cable, grids of copper tubing are sometimes embedded which circulate electrically heated hot water. Such systems are actually a form of circulating hot-water heat, and are easily as capable of being oil- or gas-fired.

INSTALLING RADIANT CEILING PANELS

Flexible and rigid panels containing electric heating grids are available for retrofitting or new construction. Some types fasten to standard framing prior to installing the finished ceiling; others are embedded in gypsum and may be installed using drywall screws or nails.

Typical radiant electric ceiling (cable-embedded)
1 Service panel
2 Heating-cable grids
3 Thermostats

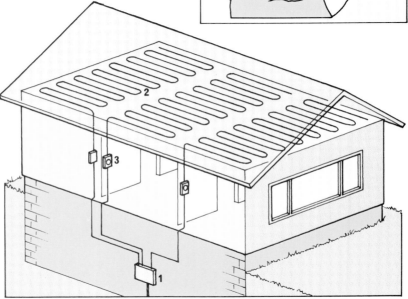

WARNING

Never make electrical connections until the power is switched off at the service panel.

Baseboard heaters

Electric baseboard heaters are the most popular form of electric heat in the U.S. Each unit contains one or more horizontal heating elements, the entire unit being thermostatically controlled. Often, units are ganged around the perimeter of the area to be heated.

Wall heaters

Individual electric wall heaters are often installed in special-use areas such as bathrooms and laundry rooms to provide supplementary or occasional heat. Most of these units include a small fans which aids in circulating heated air throughout the room.

Electric furnace

Electric furnaces are small and require no chimney or vents. They usually consist of several cooking-oven type heating elements plus a squirrel-cage blower. The furnace is linked to a duct system and operates as a form of forced hot-air heating.

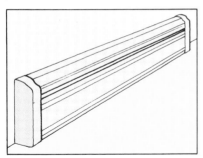

Electric baseboard heater

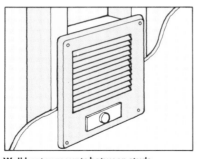

Wall heater mounts between studs

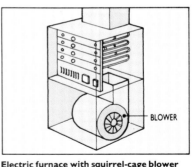

Electric furnace with squirrel-cage blower

FUEL-BURNING FURNACES

The furnace is the heart of any heating system, whether forced-air, circulating hot-water, or steam. By far the most common furnaces in use today are fuel-burners—gas- or oil-fired. Either may in turn be used to heat air or water for circulation throughout the house. Fuel-burning furnaces require regular adjustment and periodic care. If they are properly looked after, however, modern furnaces can be economical and efficient heat producers and will provide years of dependable service, often equal to the life of the house.

Oil burners

Parts of an oil furnace
1 Motor/blower
2 Combustion chamber
3 Heat exchanger
4 Chimney vent
5 On/off switch
6 Combination gauge

Oil burners spray fuel oil into a combustion chamber which is then ignited to produce heat. The two most popular burner designs are the pressure (gun) type and the vaporizing (pot) type. Pressure burners are by far the most popular furnaces in the U.S. today. Of the various designs, both high- and low-pressure types are available. The most common variety is high-pressure.

In a high-pressure oil burner, a fine spray of oil is jetted under pump pressure through a nozzle, mixed with air and then ignited by an electric spark derived from household current. Low-pressure burners are somewhat similar, the difference being that the oil and air are mixed before exiting the nozzle and pumped into the combustion chamber under far less pressure and through a much larger opening.

Vaporizing burners do not operate under pressure of any sort. The combustion area consists of an enclosed shallow pan into which oil is admitted by regulating a manually operated valve. The oil is ignited by hand or by a simple electric igniter, which causes it to vaporize, then is kept burning by a small pilot flame. Some units feature a small blower to increase the amount of air drawn into the furnace for combustion. Because vaporizing oil burners are compact and make very little noise, they are sometimes installed in kitchens or utility rooms. Pressure-type oil burners, on the other hand, are nearly always installed in basements.

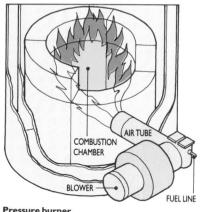

Pressure burner

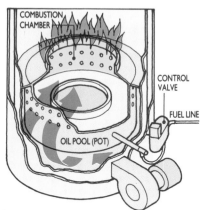

Vaporizing burner

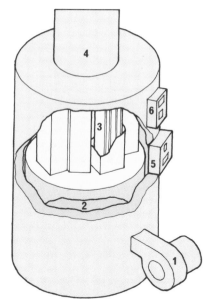

Typical oil burner (shown equipped for circulating hot water)

Gas burners

Parts of a gas furnace
1 Gas supply
2 Manual shutoff valve
3 Thermocouple
4 Pressure regulator
5 Automatic supply valve
6 Burners
7 Combustion chamber
8 Heat exchanger
9 Chimney vent

Gas burners are far simpler than oil burners and, since gas itself burns much more cleanly than fuel oil, require less regular maintenance. Still, they should be cleaned and services at least every 2 or 3 years.

Whether supplied by natural gas or liquefied petroleum (LP gas), furnaces of this type consist merely of a burner layout and gas-regulating valve. No pressurizing system is employed. The burners may be of a type that spread their flame over a large area or they may merely have multiple openings (or jets), as on a gas kitchen range.

For safety, gas burners incorporate a thermocouple device which automatically shuts off the gas-supply valve when no heat is detected in the combustion area. Should the odor of gas ever be detected near a gas burner, immediately open windows, extinguish any open flame or cigarettes, then leave the premises and telephone your gas or utility company from a phone outside the house. Do not touch any electrical switches—doing so many produce a static charge sufficient to ignite escaping gas.

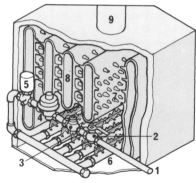

Typical gas burner (equipped for circulating hot water)

RADIATORS AND CONVECTORS

The hot water from a central heating boiler is pumped through the house along narrow pipes which may be connected to radiators or to special convector heaters that extract heat from the water and pass it out into their respective rooms.

You can feel radiant heat being emitted directly from the hot surface of an appliance, but convected heat warms the air that comes into contact with the hot surface. As the warmed air rises toward the ceiling, it allows cooler air to flow in around the appliance, and this air in turn is warmed and moves upwards. Eventually a steady but very gentle circulation of air takes place in the room, and the temperature gradually rises to the optimum set on the room thermostat.

Radiators

Ordinary radiators are made of heavy cast iron which absorbs and then radiates heat for a long time. For hot-water use only, lightweight radiators made of pressed sheet metal are sometimes available as imports. In either type, water flows in through a manually adjustable valve at one corner and then out through a return valve at the other (except in the case of most steam radiators—see box at right). A bleed valve is placed near the top of the radiator on the end opposite the inflow valve to let air out and prevent airlocks which stop the radiator from heating up properly. Cast-iron radiators are normally freestanding. However, special brackets are available to hang them from sturdy wall studs. Sheet-metal radiators are usually wall-hung.

Despite their name, radiators deliver only about half their output as radiant heat; the rest is emitted through natural convection as the surrounding air comes into contact with the hot surfaces of the radiators.

Radiators come in a wide range of sizes, and the larger they are, the greater their heat output. Maximum efficiency dictates that radiators be fully exposed in a room, however, not recessed or covered by a vented housing. Nor should they be painted or hidden behind furniture or drapes. A better way to deal with unsightly radiators in an interior decor is to replace them with less obtrusive convectors.

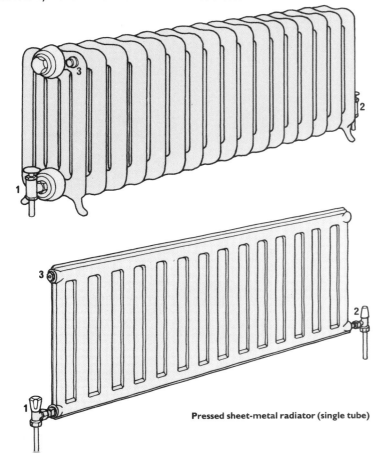

Pressed sheet-metal radiator (single tube)

STEAM UNITS

Cast-iron radiators designed for steam heat look similar to those used for hot water and, with minor modifications, are interchangeable. In most cases steam radiators have only one pipe connecting them with the main supply line; water that has condensed as the steam gives up its heat makes its way back to the boiler the way it came. The other difference is that in place of a bleed valve, steam radiators have an automatic vent built into the end of the radiator opposite the inlet pipe. The vent permits air to bleed out automatically as the radiator fills with steam. However, the steam itself does not escape.

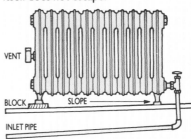

To cure noisy radiator, slope end toward inlet. Pipe should slope toward furnace.

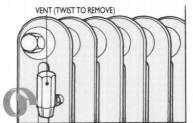

If radiator heats unevenly, remove vent and listen for escaping air. Replace with new vent.

Maintenance
Many steam radiators produce knocking and banging noises as they heat up. The sound is actually made by water which has become trapped striking the walls of the radiator or piping as the steam seeks to get past. To cure the condition, make sure that the radiator slopes slightly downward toward the inlet pipe and the pipe itself slopes downward toward the furnace.

If the radiator will not heat properly all the way across, suspect a blocked vent. Air that cannot escape prevents steam from diffusing throughout the radiator. Shortly after turning on the heat, unscrew the vent and remove it. As the steam rises, air should escape from the hole, followed by steam, indicating that the radiator itself is functioning properly. Purchase a new vent and install it in place of the old one.

SEE ALSO

Details for: ▷

Steam heat	56
Convectors	60
Thermostats	61
Circulating pumps	67

Cast-iron radiator
1 The manual valve controls inflow.
2 The return valve controls outflow to keep the radiator hot.
3 The bleed valve is to disperse airlocks.

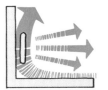

Heat emission
The large arrow shows the flow of air during convection. The small arrows indicate how heat radiates from the radiator's surface.

CONVECTORS

You can install convector heaters in your circulating hot-water heating system to replace conventional cast-iron radiators.

Unlike radiators, convector heaters emit none of their heat in the form of direct radiation. The hot water from the boiler passes through a finned pipe inside the heater, the fins absorbing the heat and transferring it to the air surrounding them. The warmed air escapes through an opening at the top of the appliance and at the same time cool air is drawn in through the open bottom, to be warmed in turn.

Most convector heaters have a damper that can be set to control the airflow. Some are designed for inconspicuous mounting at baseboard level. An advance is a fan-assisted type in which the airflow over the heating fins is forced, making for fast room heating.

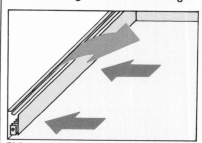
Rising warm air draws cool air in below

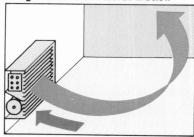
Airflow by fan-assisted convection

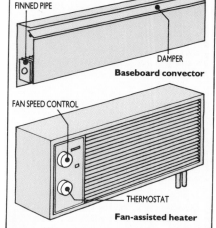
FINNED PIPE

DAMPER
Baseboard convector

FAN SPEED CONTROL

THERMOSTAT
Fan-assisted heater

PLACING YOUR HEATERS

At one time central heating radiators or convectors were nearly always placed under windows to balance the chill of the panes and cut the drafts caused by warmed air cooling against them. If double glazing deals with both of these problems you can place radiators or convectors with an eye to maximum comfort— but keep the other on the length and consequent cost of pipe runs.

Convenience and cost

Your radiators and convector heaters can be positioned anywhere that's convenient, that suits the shape of the rooms and that keeps costly pipe runs to a reasonable minimum. Probably you'll want to take all of these considerations into account.

While double glazing means that the appliances can be sited elsewhere than under windows, there is a slight drawback to placing them against walls. The warm air rising from them will tend to discolor the paint or wallpaper above. You can guard against this by fitting radiator shelves immediately above them to direct the warm air clear of the walls.

Never hang curtains or stand furniture in front of radiators or convectors. They will absorb radiated heat, and curtains will trap convected heat between themselves and the walls. While convectors radiate almost no heat, you should never obstruct warm air leaving the appliance nor cool air being drawn into it.

A room's shape can affect the siting of appliances and perhaps their number. For example, you cannot heat a large L-shaped room from a radiator in its short end. A heating installer can work out a combination of appliances—and their positions—to heat a room properly.

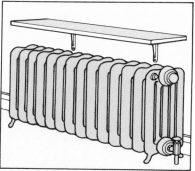
A shelf directs warm air clear of the wall

Selecting the size of heaters

A house loses heat constantly— whenever a door or window is opened; because of cold drafts; through the fabric of the doors, windows, walls, floors, ceilings and roof. To work out the heating needs of the rooms in a house, the designer has to take into account the rate at which they lose heat. This varies with the materials and construction of the walls, floors and ceilings. For example, heat is lost more quickly through a solid brick wall than through an insulated wood-frame wall. Also, the temperature on the other side of walls, floors and ceilings will come into the equation.

The designer also needs to know the temperature to which each room must be heated, and there are standard levels for particular rooms. The designer's calculations will produce a heating requirement for each room, expressed in kilowatts; the next step is to select radiators or convectors of appropriate outputs. Then all the heat output figures are totaled to give the output required from the boiler.

When you install your central heating, be sure to choose radiators and convector heaters that meet the standards approved by your local building inspector.

IDEAL ROOM TEMPERATURES

A central heating designer/installer aims at a system that will heat rooms to standard or customer-specified temperatures. Such a design might look like this:

ROOM	TEMPERATURE
Living room	70°F
Dining room	70°F
Kitchen	60°F
Hall/landing	65°F
Bedroom	60°F
Bathroom	72°F

CENTRAL HEATING CONTROLS

A range of automatic control systems and devices for circulating central heating can, if used sensibly, enable you to make useful savings in running costs. They can ensure that your system never "burns up money" by producing unwanted heat.

Three basic devices

While considerable sophistication is now available in automatic control, the systems can be divided into three main types: temperature controllers (thermostats), automatic on-off switches (timers and programmers) and heating circuit controllers (zone valves).

These devices can be used individually or in combination to provide a very high level of control.

It must be added that automatic controls are really effective only with gas- or oil-fired boilers, which can be switched on and off at will. Linked to coal or wood systems, which take time to react to controls, the systems will be less effective and can be dangerous.

ZONE CONTROL VALVES

It is not often that all the rooms in a house are in use at once. During the day it is normal for the upstairs rooms to be unused for long periods, and to heat them permanently would be very wasteful. To avoid such waste you can divide your central heating system into circuits, or zones—the usual ones being upstairs and downstairs—and heat those areas only when it's necessary.

Control is provided by motorized valves linked to a timer or programmer that directs the flow of heating water through preselected pipes at preselected times. Alternatively, zone valves can be used to provide zone temperature control by being linked to individual zone thermostats.

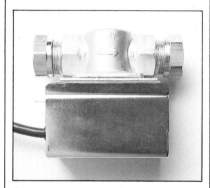

A motorized zone control valve

Thermostats

All steam and hot-water systems incorporate thermostats to prevent overheating. A gas- or oil-fired tank will have one that can be set to alter heat output by switching the unit on and off. Another, called an aquastat, can be set to monitor the temperature of the water circulating through the pipes.

Room thermostats are common forms of central heating control, often the only ones fitted. They are placed in rooms where temperatures usually remain fairly stable, and work on the assumption that any rise or drop in room temperature will be matched by similar ones throughout the house. Room thermostats control temperatures through simple on-off switching of the heating unit—or its pump if a boiler must run constantly to provide a constant supply of domestic hot water.

The room thermostat's drawback is that it can make no allowance for local temperature changes in other rooms caused, for example, by the sun shining through a window or a separate heater being switched on. Much more sophisticated temperature control is provided by thermostatic radiator valves, which can be fitted to radiators instead of the standard manually operated inlet valves. Temperature sensors open and close them, varying heat output to maintain the desired temperatures in individual rooms.

Thermostatic radiator valves need not be fitted in every room. You can use one to reduce the heat in a kitchen or reduce the temperature in a bathroom while using a room thermostat to regulate the temperature in the rest of the house or separate zones.

Other available thermostatic controls include devices for regulating the temperature of domestic hot water and for giving frost protection to a unit switched off during winter holidays.

Timers and programmers

You can save a lot in running costs by ensuring that your heating system is not working when you don't need it—while you're out, for instance, or while you sleep. A timer can be set to switch the system on and off to suit the regular comings and goings of the family. It can switch on and warm the house just before you get up, then off again just before you leave for work, on again when you come home, and so on.

The simpler timers offer two "on" and two "off" settings which are repeated daily, though a manual override allows variations for weekends and such. More sophisticated programmable versions offer a number of on-off options—even a different one for each day of the week—as well as control of hot water.

Timer

Room thermostat

Programmable

Thermostatic radiator valve

Heating controls
There are several ways to control the temperature:
1 Programmable type controls boiler and pump.
2 A timer is used to control a zone valve. It can be used to regulate boiler and pump.
3 Room thermometer controls pump or a zone valve.
4 A nonelectrical radiator valve controls an individual heater.

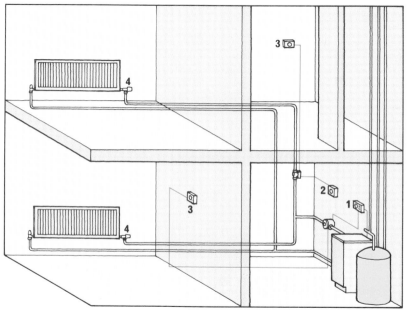

DRAINING A HOT-WATER SYSTEM

Circulating hot-water systems rarely need complete draining. However, if the water has become overly contaminated, or if a component fails and must be removed for replacement, by following the procedures outlined here, the task of removing the water from the boiler and piping can be done fairly easily.

Steps in draining hot-water system
1 Extinguish furnace
2 Shut off water supply line
3 Open draincock
4 Open radiator bleed valves
5 Drain expansion tank

Draining the system

The most frequent reason for draining a hot-water system is excessive rust in the circulating water. Each year, a small amount of water should be drained from the boiler into a clear glass via the draincock usually located near the base of the unit. Should the water appear unusually cloudy, draining the system, flushing it clean, and then refilling it is in order.

Begin by turning off the furnace. Remember that with a gas furnace this means also turning off the pilot flame and main gas inlet. Next, turn off the water supply line to the boiler. Connect a garden hose to the draincock and lead the other end of the hose to a floor drain. If no drain is present, position a bucket beneath the draincock. After waiting until you are certain the water has cooled sufficiently, open the draincock and let the water drain out. While it is draining, open the bleed valves on the radiators in the house to avoid creating a partial vacuum which will prevent complete drainage. At the same time, drain the expansion tank (see below).

Flushing

To rid the boiler of accumulated rust and sediment, leave the draincock open after the water stops flowing, then reopen the water supply line to admit fresh water into the system. When the water runs clear, close the draincock and let the boiler fill.

REFILLING THE SYSTEM

When you are ready to refill the boiler, close the draincock used for flushing the system. If you wish to add a commercial rust inhibitor to forestall further corrosion in the pipes, do so by closing off the water supply line, removing the pressure relief valve from the boiler tank and pouring the recommended amount of inhibitor (see manufacturer's instructions) into the hole, then replacing the valve. Reopen the water supply line. Wait until the boiler fills, then turn on the furnace. Reclose the radiator bleed valves when you hear water rising in the radiators. Wait several hours with the system running, then bleed all the radiators.

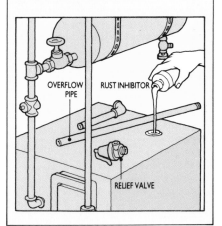

Draining the expansion tank

Conventional expansion tanks are merely cylinders partially filled with water. As the water in the circulating system heats and expands, air present in the tank is compressed, relieving the excess system pressure created and also preventing the hot water from turning to steam. Over a period of time, however, most tanks gradually accumulate too much water, which forces the air out and thus prevents the tank from functioning properly. The solution is to drain the tank, an operation that must be done also if the entire system is to be drained.

To drain a conventional expansion tank, attach a garden hose or a bucket to the draincock, usually located on the underside of the tank. Close off the inlet pipe leading to the tank, then open the draincock. If no inlet valve is present, you must drain the entire system in order to drain the tank as well.

Diaphragm expansion tanks

Some expansion tanks physically separate air and water by means of a rubber partition (diaphragm) which divides the tank into two chambers. Instead of draining the water from such tanks, they are periodically recharged with air. To do so, check the air pressure in the chamber using an ordinary tire-pressure gauge attached to the recharge valve, usually located on the underside of the tank. Then use a bicycle pump to add air until gauge readings match the tank's recommended pressure. NOTE: A diaphragm tank which requires even moderately frequent recharging may be leaking and should be replaced.

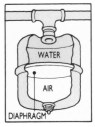

Diaphragm expansion tanks never need draining
To recharge air chamber, use a bicycle pump.

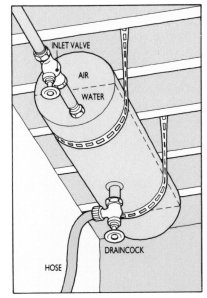

Conventional expansion tank

Routine cleaning, maintenance and adjustment will give your central heating furnace a longer life and prolong its efficiency.

Both gas and oil-fired furnaces should be serviced once a year by qualified service persons, but you can do a certain amount of cleaning and tuning up yourself.

Servicing oil burners

Here are several maintenance chores you can perform which do not involve making adjustments to the combustion components of the furnace. Unless you are skilled and have the proper testing instruments, leave those to a service professional. Before starting any maintenance task, turn the furnace completely off.

Maintaining gas burners

Because gas furnaces involve less complicated equipment and technology than oil burners (combining gas and air for combustion is much easier than combining fuel oil and air), frequent maintenance of gas furnaces is less necessary than with oil-fired units. Inspection of the flame and pilot mechanisms should be carried out yearly by a professional service technician, of course. Cleaning—a task suitable to do-it-yourselfers—normally need take place only every 2 or 3 years. Before beginning any maintenance task, be certain to turn off the main gas supply line, pilot light, and burners.

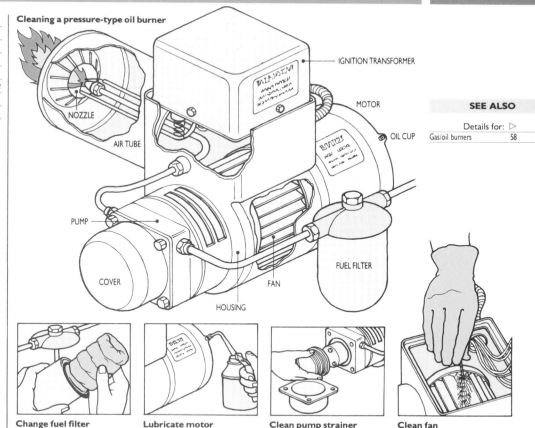

Cleaning a pressure-type oil burner

IGNITION TRANSFORMER

MOTOR

OIL CUP

NOZZLE

AIR TUBE

PUMP

COVER

FAN

HOUSING

FUEL FILTER

SEE ALSO

Details for: ▷
Gas/oil burners   58

Change fuel filter
Place pan beneath filter area. Unscrew cup. Remove and replace cartridge and gasket.

Lubricate motor
Locate oil cups at each end of motor (if none are present, motor is self-lubricating and requires no treatment). Squirt 3 to 6 drops SAE 10W nondetergent electric motor oil in each cup. Do not overlubricate.

Clean pump strainer
Remove pump cover and gasket. Soak strainer in solvent, then brush clean with toothbrush. Replace, using new gasket. (Note: Some pumps do not have strainers.)

Clean fan
Unbolt transformer and swing aside to access fan. Clean blades with bottle brush or lint-free cloth attached to popsicle stick; wipe interior of housing with rag.

PROFESSIONAL SERVICE PLANS

The very high efficiency of modern gas and oil furnaces largely depends on their being regularly checked and serviced. It should be done annually, and because the mechanisms involved are so complex, the work should be done by qualified service persons.

For either type of furnace you can enter into a contract—with either the original installer or the fuel supplier—for regular maintenance.

Gas-fired installations
Many gas utilities offer a choice of several service arrangements for gas furnaces. These cover their own installations, but they can often be arranged for systems put in by other installers on condition that the utility inspect the installation before writing the contract.

The simplest maintenance contract plan provides for an annual check and adjustment of the furnace. If any repairs are found to be necessary, either at the time of the regular check or at other times during the year, the labor and the required parts will be charged separately. But for an extra fee it is possible to have both free labor and free parts for furnace repairs at any time of the year. Most utilities will also extend the arrangement to include a check of the whole heating system at the same time as the furnace is being checked.

It may be that your own installer can offer you a similar choice of service plans. The best course is to compare the charges and decide which gives the best value for your money.

Oil-fired installations
The installers of oil-fired central heating systems and the suppliers of fuel oil offer service plans similar to those outlined above for gas-fired systems. The choice of plans ranges from the simple checkup each year to complete coverage for new parts and labor if and when any repairs should become necessary.

Again, as with the plans for gas, it is wisest to shop around and make a comparison of the various services offered and the changes for them.

BLEEDING RADIATORS AND FITTING A BLEED VALVE

If a radiator feels cooler at the top than at the bottom, it's likely that a pocket of air has formed in it and is stopping full circulation of the water. Getting the air out—"bleeding"—is a simple matter.

Opening a bleed valve

Bleed radiators with the circulator running. Each radiator has a bleed valve at one of its top corners, near the end opposite the water inlet. Usually the valve is slotted for a screwdriver, but on many new models the valve is a square-section shank in the center of the round blanking plug. You should have been given a key to fit these shanks by the installer, but if you weren't, you can buy one at a plumbing supply store.

Use the key to turn the shank of the valve counterclockwise about a quarter of a turn. It shouldn't be necessary turn it further, but have a small container handy to catch any spurting water if you do open the valve too far.

You will hear a hissing sound as the air escapes. Keep the key on the shank of the valve and when the hissing stops and the first dribble of water appears, close the valve tightly.

In no circumstances must you be tempted to open the valve any more than is needed to let the air out, or to remove it completely, as this will produce a deluge of water.

Dispensing the air pocket in a radiator

Radiator key
Keep your radiator key in a handy place where you can find it on short notice. There is no substitute for it.

Fitting an automatic bleed valve

If you find yourself having to bleed one particular radiator regularly it will save you trouble if you replace its bleed valve with an automatic one that will allow air to escape but not water.

First drain the water from the system, then use your bleed-valve key to unscrew the old valve completely out of the blanking plug (1). Wind some Teflon tape around the threads of the automatic valve (2) and screw it finger-tight into the blanking plug (3).

Refill the system; if any water appears around the threads of the new valve, tighten it further with a wrench (4).

If, when the system is going again, the radiator still feels cool on top, it may be that a larger amount of air has collected than the bleed valve can cope with. In this case, unscrew the valve until you hear air hissing out. Tighten it again when the hissing stops and the first trickle of water appears.

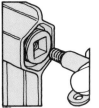

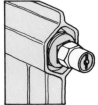

1 Unscrew old valve   **2 Tape the new one**   **3 Screw it finger-tight**   **4 Stop any leak in use**

HOW TO REMOVE A RADIATOR

You can remove an individual radiator while the wall behind it is decorated without having to drain the whole system. You simply close the valves at the ends of the radiator, drain it and then remove it.

Shut off both valves, turning the shank of the return valve clockwise with a key or an adjustable wrench (1). Note the number of turns needed to close it so you can reopen it by the same number of turns later.

Make sure that you have plenty of rags for mopping up spillage, also a jug and a large bowl. As the water in the radiator will be very dirty, you should also roll back the floorcovering before you start, if possible.

Unscrew the capnut that holds one of the valves to the adapter in the end

of the radiator (2). Hold the jug under the joint and open the bleed valve slowly (3) to let the water drain out. Transfer the water from jug to bowl and keep going until no more can be drained.

Unscrew the capnut that holds the other valve on the radiator. If the radiator is not freestanding, lift it free from its wall brackets (4) and drain any remaining water into the bowl. Unscrew the wall brackets to decorate.

To replace the radiator after decorating, screw the brackets back in place, hang the radiator on them and tighten the capnuts on both valves. Close the bleed valve and open both radiator valves. Adjust the return valve by the same number of turns as you used to close it. Finally, use the bleed valve to release any trapped air.

1 Close the valve   **2 Unscrew capnut**

3 Open bleed valve

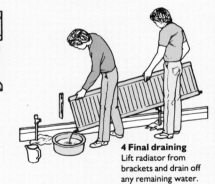

4 Final draining
Lift radiator from brackets and drain off any remaining water.

RADIATOR VALVES

Curing a leaking radiator valve

If an inlet valve or return valve on a radiator seems to be leaking, its most likely that one of the capnuts that secures it to the water pipe and to the radiator's valve adapter needs some tightening up.

Tighten the suspect capnut with an adjustable wrench while you hold the body of the valve with a pipe wrench to prevent it from moving. If this doesn't work, the valve will have to be replaced.

If the leak seems to be from the valve adapter in the radiator, the joint will have to be repaired in the same way as when a new valve is fitted (see below). If the leak seems to be from the valve spindle, tighten the gland nut (see left) or replace the valve.

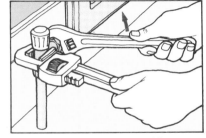

Grip leaky valve with wrench to tighten capnut

Replacing a worn or damaged valve

Be sure that the new valve is exactly like the old one, or it may not align with the water pipe. Drain the heating system and lay rags under the valve to catch any remaining dregs that may come out.

Hold the body of the valve with a pipe wrench and use an adjustable wrench to unscrew the capnuts that hold the valve to the water pipe and to the adapter in the end of the radiator (1). Lift the valve from the end of the pipe (2). If the valve being replaced is a return valve, don't remove it before you have closed it, counting the number of turns needed so that you can open the new valve by the same number to balance the radiator.

Unscrew the adapter from the radiator (3). You may manage this with an adjustable wrench or you may need an Allen wrench, depending on the type of adapter.

Fitting the new one

Ensure that the threads in the end of the radiator are clean and wind Teflon tape four or five times around the thread of the new valve's adapter, then screw it into the end of the radiator by hand and tighten it a further 1½ turns with an adjustable or Allen wrench.

Slide the valve capnut and a new ferrule over the end of the water pipe and fit the valve to the end of the pipe (4), but don't tighten the capnuts at this stage. First align the valve body with the adapter and tighten the capnut that holds them together (5). Hold the valve body firm with a wrench while you do this. Now tighten the capnut that holds the valve to the water pipe (6).

Finally, refill the system, check for leaks at the joints of the new valves and tighten the capnuts some more if that seems necessary.

DEALING WITH A JAMMED FERRULE

If the ferrule is jammed onto the pipe, cut the pipe off below floorboard level and make up a new section. Join it to the old pipe by means of either a soldered joint or a compression joint.

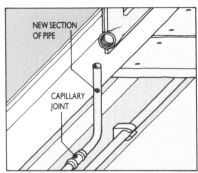

New section replaces pipe with jammed ferrule

Tightening gland nut
Tighten the gland nut with a wrench to stop a leak from a radiator valve spindle. If the leak persists, replace the valve (see right).

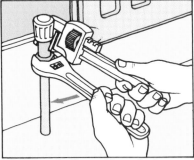

1 Hold the valve firm and loosen both capnuts

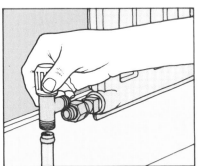

2 Unscrew the capnuts and lift the valve out

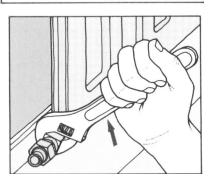

3 Remove the valve adapter from the radiator

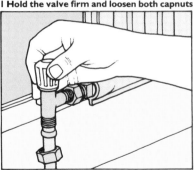

4 Fit new adapter, then fit new valve to pipe

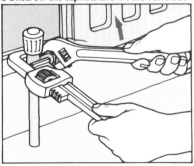

5 Connect valve to adapter and tighten capnut

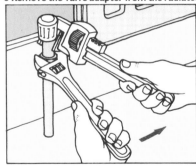

6 Tighten capnut holding valve to water pipe

REPLACING A CORRODED OR DAMAGED RADIATOR

Try to obtain a new radiator of exactly the same model as the one you wish to replace. This will make the job easier.

Simple replacement

Drain and remove the old radiator. With it clear of the wall, unscrew the valve adapters from the bottom with an adjustable wrench or, if necessary, an Allen wrench. Unscrew the bleed valve with its key, and then the two blanking plugs from the top of the radiator, using a square or hexagonal Allen wrench (1).

With steel wool, clean up the threads of both adapters and both blanking plugs (2), then wind four or five turns of Teflon tape around the threads (3). Screw them into the new radiator and the bleed valve into the blanking plug.

Position the new radiator and connect the valves to their adapters. Open the valves and fill and bleed the radiators.

1 Taking the measurements
Measure the positions of the radiator brackets and transfer the measurements to the wall. Double-check the results to make sure the radiator is equidistant between the two pipes projecting from the floor.

1 Removing the plug
Use an Allen wrench to unscrew the blanking plug at each end of the radiator.

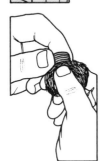

2 Cleaning the threads
Use steel wool to clean any corrosion from the threads of both blanking plugs and valve adapters.

3 Taping the threads
Make the threaded joints watertight by wrapping Teflon tape several times around each component before screwing them into the new radiator.

Replacement with a different pattern radiator

Rather more work is involved in the replacement if you can't get a radiator of the same pattern as the old one. If it is not freestanding, you'll have to fit new wall brackets. Possibly you'll also have to alter the water pipes.

Drain the system. Then for a wall-hung unit, take the old brackets off the wall. Lay the new radiator face down on the floor and slide one of its brackets onto the hangers welded to the back of the radiator. Measure from the top of the bracket to the bottom of the radiator, add 4 or 5 inches for clearance under the radiator, then mark a horizontal line on the wall that distance from the floor. Now measure the distance between the centers of the radiator hangers and make two marks

on the horizontal line that distance apart and at equal distances from the two water pipes (1).

Line up the brackets with the pencil marks, mark their mounting screw holes, drill and plug the holes and install the brackets in place (2).

Lift up the floorboards below the radiator and cut off the vertical portions of the inlet and return pipes. Connect the valves to the radiator and hang it on its brackets. Slip a short length of pipe into each valve as a guide for any further trimming of the pipes. Connect these lengths to the original pipes (3) with soldered or compression fittings, then connect the new pipes to the valves. Refill the system and check for leaks.

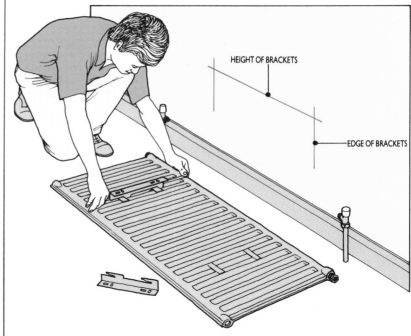

HEIGHT OF BRACKETS

EDGE OF BRACKETS

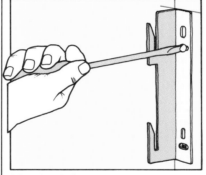

2 Securing the brackets
Screw the mounting brackets to the wall
Make sure they are on the right side of the line.

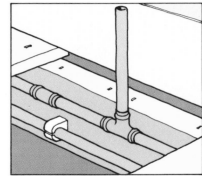

3 Connecting up
Connect the new section of pipework.
The vertical pipe aligns with the radiator valve.

CIRCULATING PUMPS

Forced hot-water heating depends on a steady cycle of hot water from boiler to radiators and back to the boiler for *reheating. This is the pump's job. A faulty pump means poor circulation. A failed pump means no circulation.*

Bleeding the pump

If your radiators don't seem to be warming up, though you can hear or feel the pump running, it's likely that an airlock has formed in the pump and its impeller is spinning in air. The air must be bled from the pump, a job that's done in the same way as bleeding a radiator. You'll find a screw-in valve for the purpose in the pump's outer casing.

The valve's position varies with the different makes, but it is usually marked.

Switch off the pump, have a jug or jam jar handy to catch any water spillage and open the valve with a screwdriver or vent key. Open it only slightly, until you hear air hissing out. When the hissing stops and a drop of water appears, close the valve fully.

Open the bleed valve with a screwdriver

Adjusting the pump

Central heating pumps are of two kinds: fixed-head and variable-head. Fixed-head units run at a single speed, forcing the water round the system at a fixed rate. Variable-head pumps can be adjusted to run at different speeds, circulating the water at different rates.

When a variable-head pump is fitted as part of a central heating system, the installer adjusts its speed after balancing all the radiators so that each room reaches its "design temperature." If you find that your rooms are not as warm as you would like, though you have opened the inlet radiator valves fully, you can adjust the pump speed. But first check that all radiators show the same temperature drop between their inlets and outlets. You can get clip-on thermometers for the job from a

plumber's supply store. You will need a pair of them.

Clip one thermometer to the feed pipe just below the radiator valve and the other to the return pipe below its valve (**1**). The difference between the temperatures registered by the two should be about 20 degrees F. If it is not, uncover the return valve and close it further (to increase the difference) or open it more (to reduce the difference).

Having balanced the radiators you can now adjust the pump. Switch it off and then turn the speed adjustment up (**2**), one step at a time, until you are getting the overall temperatures you want. You may be able to work the adjustment by hand or you may need some special tool, such as an Allen wrench, depending on the make and model of your pump.

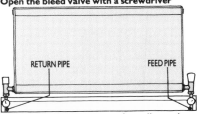

1 Clip the thermometers to the radiator pipes

RETURN PIPE   FEED PIPE

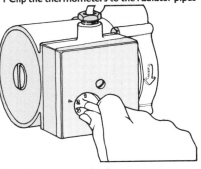

2 Adjust pump speed to increase temperature

Replacing a worn pump

If you have to replace your circulating pump, be quite sure that the one you buy is of exactly the same make and model as the old one, or seek the advice of a professional installer.

Turn off the boiler and close the isolating valves, on each side of the pump. If there are no isolating valves you will have to drain down the whole system.

Identify the electrical circuit that controls the circulating pump and trip the circuit breaker for that circuit at the service panel, then take the cover plate off the pump and disconnect its wiring (**1**).

Have a bowl or bucket at hand to catch any water left in the pump, also some old rags for any mopping up that you may have to do. Undo the retaining nuts that hold the pump to the valves or the pipework with an adjustable wrench

(**2**) and catch the water as it flows out.

Remove the pump and fit the new one in its place, taking care to fit correctly any sealing washers that are provided (**3**), then tighten up the retaining nuts.

Take the cover plate off the new pump, feed in the electrical cable, connect the wires to the pump's terminals (**4**) and replace the cover plate. If the pump is of the variable-head type (see above), set the speed control to the speed indicated on the old pump.

Open both isolating valves—or if the system has been drained, refill it—check the pump connections for leaks and tighten them if necessary. Then open the pump's bleed valve to release any air that may have become trapped in it. Finally, restore power to the circuit and test the pump.

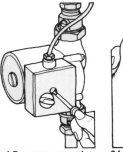

1 Remove cover plate   **2 Undo connecting nuts**

3 Attach new pump   **4 Connect power cable**

WARNING

Never make electrical connections until the power is switched off at the service panel.

CONTROL VALVES

The control valves are vital to the effective working of a modern central heating system, for it is through them that the various timers and thermostats are able to adjust the levels of heating precisely to the programmed requirements you set for comfort.

Worn or faulty control valves can seriously affect the reliability of the system and should be replaced promptly.

WARNING

Never make electrical connections until the power is switched off at the service panel.

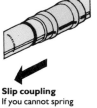

Slip coupling
If you cannot spring pipework to locate a conventional soldered joint, use a slip coupling which is free to slide along the pipe to cover the junction.

Far left
Two-port control valve
A two-port valve seals off a section of pipework when water within that section has reached the required temperature.

Left
Three-port control valve
This type of valve can independently isolate central heating or hot-water circuits.

68

Replacing a faulty valve

First ensure that the replacement valve that you buy is of exactly the same pattern as the faulty one, or seek professional advice.

Drain down the system, then identify the electrical circuit that services the central heating controls and turn off its power at the service panel.

The electrical cable from the valve will be connected to the terminals of a nearby junction box, which will also be linked to the heating system's other controls. Take the cover off the junction box and disconnect the wiring for the valve. As you do this you should carefully note the connections so as to make reconnection easier.

You will probably not be able to take the old valve out of the pipe run by simply unscrewing its capnuts, as you won't be able to pull the ends of the pipe free of their sockets in the valve. Instead, cut through the pipe on each side of the valve (**1**) and take out the section, complete with valve, then make up two pieces of pipe to fit on either side of the new control valve.

Assemble the valve with its ferrules and capnuts, pipes and joints, but only loosely at first, and fit the assembly into the pipe run (**2**).

There should be enough play in the joints to allow the assembly to be sprung into place, and this may be helped if the original pipes on each side of the valve position are first freed from their clips.

When the pipes and valve are in place, connect them to the original pipework with compression or soldered joints, then tighten the valve capnuts. Hold the body of the valve with a second wrench to prevent it from turning (**3**).

Reconnect the cable to the terminals of the junction box and replace its cover plate, then restore power to the circuit.

Refill the heating system and check the working of the valve by adjusting the timer or thermostat that controls it.

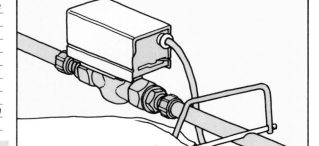

1 Removing the valve
If you cannot disconnect the valve, cut through the pipe on each side of it with a hacksaw.

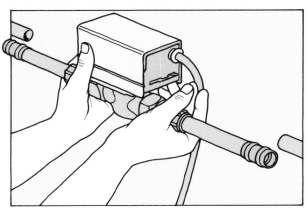

2 Fitting new valve
With the new valve connected to short sections of pipe, spring the assembly into the pipe run.

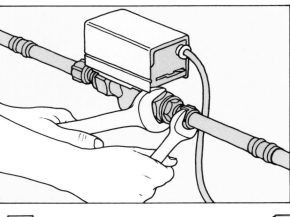

3 Closing the joints
Having connected the pipework, tighten the valve capnuts on each side with a pair of wrenches.

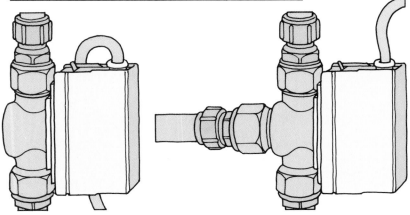

SYMPTOM

Hissing or banging sounds from boiler or pipework.

Overheating caused by:

- Blocked chimney (with coal furnace).
  Check flueway for substantial soot fall. Sweep chimney.

- Heavy scale deposits in system due to hard water.
  Shut down boiler and pump. Have a specialist treat system with a descaler, then drain, flush and refill system.

- Faulty boiler thermostat.
  Shut down boiler. Leave pump working to circulate water around system and cool it quickly. With system cool, operate boiler thermostat control. If you do not hear a clicking sound, call in a repair person.

- Incorrectly sloping pipework (with steam systems).
  Check to see that radiators and pipe runs carrying steam slope downward at all points, so that water can travel freely back to boiler. Insert wooden wedges under improperly sloping radiators. Rehang improperly pitched pipe. Check results using carpenter's level.

- Circulating pump not working (with coal furnace).
  Shut down boiler, then check that the pump is switched on. If pump won't work, turn off power and check wired connections to it. If pump seems to be running but outlet pipe is cool, check for airlock by using bleed screw. If pump is still not working, shut it down, drain system, remove pump and check it for blockage. Clean pump or replace it if necessary.

SYMPTOM

Pressure-relief valve on circulating hot-water boiler opens, sending water through overflow pipe.

- Excess pressure in piping caused by malfunctioning expansion tank.
  Drain conventional tank by first closing inlet valve leading to tank, then opening draincock on underside of tank. For tanks with diaphragm-design, recharge air chamber using bicycle pump, or replace tank.

SYMPTOM

All radiators remain cool though boiler is operating normally.

- Pump not pumping.
  Check that pump is working by feeling for motor vibration or by listening. If pump is running, check for airlock by operating bleed valve. If this has no effect, the pump outlet may be blocked. Switch off boiler and pump; remove pump and clean or replace as necessary.

- Pump's thermostat or timer is set incorrectly or is faulty.
  Check thermostat or timer setting and reset if necessary. If this makes no difference, switch off power and check wiring connections. If connections are in good order, call in repair person.

SYMPTOM

Radiators in one part of the house do not warm up.

- Timer or thermostat which controls zone valve not set properly or faulty.
  Check timer or thermostat setting and reset if necessary. If this has no effect, switch off the power supply and check wired connections. If this makes no difference, call in repair person.

- Zone valve itself faulty.
  Drain system and replace valve.

SYMPTOM

Single radiator does not warm up.

- Manual inlet valve closed.
  Check setting of valve and open it if that is necessary.

- Thermostatic radiator valve not set properly or faulty.
  Check setting of valve and reset it if necessary. If this has no effect, drain radiator and replace valve.

- Return valve not set properly.
  Remove return valve cover and adjust valve setting until radiator seems as warm as those in adjacent rooms. Have valve properly balanced next service.

- Inlet/outlet blocked by corrosion.
  Close inlet and return valves, remove radiator and flush out and refit or replace as necessary.

SYMPTOM

Area at top of radiator stays cool while the bottom is warm.

- Airlock at top of radiator preventing water from circulating fully.
  Operate bleed valve to release the trapped air.

SYMPTOM

Cool patch in center of radiator while top and ends are warm.

- Heavy deposits of corrosion at bottom of radiator are restricting circulation of water.
  Close inlet and return valves, remove radiator, flush out, then refit or replace as necessary.

SYMPTOM

Water leaking from system.

- Loose pipe unions at joints, pump connections, boiler connections, etc.
  Switch off boiler and turn furnace off completely—switch off pump and tighten leaking joints. If this has no effect, drain the system and remake joints completely.

- Split or punctured pipes.
  Wrap damage in rags temporarily, switch off boiler and pump and make a temporary repair with hose or commercial leak sealant. Drain system and fit new pipe.

SYMPTOM

Boiler not working.

- Thermostat set too low.
  Check that room or boiler thermostat is set correctly.
- Timer or programmer not working.
  Check that unit is switched on and set correctly. Have it replaced if the fault persists.
- Pilot light goes out.
  Relight a gas boiler pilot light following the manufacturer's instructions, which are usually found on the back of the boiler's front panel. If the pilot fails to ignite after second try, have the unit replaced.

BUILDING CODES, PERMITS AND OTHER LEGAL CONSIDERATIONS

OFFICIAL PERMISSION

Before starting certain building projects, it is necessary to obtain official permission from local authorities. Depending on the type of project and local regulations, you may need to apply for a building permit, zoning variance or a certificate of appropriateness. In some cases, you may need all three.

Building codes

Building codes address nearly every detail of building construction from the acceptable recipes for concrete used in the foundation to the permissible fire rating of the roof finish material—and many features in between. Partly because codes attempt to be as comprehensive as possible and also because they must address different concerns in varied locales, they are very lengthy, complex and lack uniformity from region to region. A further complication is that many new building products become available each year that are not accounted for in existing codes. Model codes promulgated by four major organizations are widely used for reference throughout the United States.

The Uniform Building Code, published by the International Conference of Building Officials, is perhaps the most widely accepted code. ICBO republishes the entire code every three years and publishes revisions annually. A short form of the Uniform Building Code covering buildings with less than three stories and less than 6,000 square feet of ground floor area is available—easier for home builders, and remodelers' reference.

The BOCA-Basic Building Code, issued by the Building Officials and Code Administrators International, Inc., is also widely used. An abridged form designed for residential construction, which includes plumbing and wiring standards, is available.

A third model code, prepared under the supervision of the American Insurance Association and known as the National Building Code, serves as the basis for codes adopted by many communities. It, too, is available in a short form for matters related to home construction.

The Standard Building Code is published by the Southern Building

Code Congress International, Inc. It addresses conditions and problems prevalent in the southern United States.

While it is likely that one of the model codes named above serves as the basis for the building code in your community, municipal governments and states frequently add standards and restrictions. It is your local building department that ultimately decides what is acceptable and what is not. Consult that agency if a code question should arise.

Building codes are primarily designed for the safety of the building occupants and the general welfare of the community at large. It is wise to follow *all* practices outlined by the prevailing code in your area.

Building permits

A building permit is generally required for new construction, remodeling projects that involve structural changes or additions, and major demolition projects. In some locales it may be necessary to obtain a building permit for constructing in-ground pools, and you may need a building permit or rigger's license to erect scaffolding as an adjunct to nonstructural work on a house.

To obtain a building permit, you must file forms prescribed by your local building department that answer questions about the proposed site and project. In addition, it is necessary to file a complete set of drawings of the project along with detailed specifications. A complete set normally includes a plot plan or survey, foundation plan, floor plans, wall sections and electrical, plumbing and mechanical plans. Building permit fees are usually assessed based on the estimated cost of construction and records of the application are usually passed along to the local tax department for reassessment of the property value.

At the time you apply for a building permit, you may be advised of other applications for official permission that are required. For example, you may need to apply to the county health department concerning projects that may affect sewerage facilities and natural water supplies. It is important to arrange inspections in a timely way since finish stages cannot proceed until the structural, electrical, plumbing and mechanical work are approved.

Anyone may apply for a building

permit, but it is usually best to have an architect or contractor file in your behalf, even if you plan to do the work yourself.

Zoning restrictions

Even for projects that do not require a building permit, local zoning regulations may limit the scope and nature of the construction permitted. Whereas building codes and permit regulations relate to a building itself, zoning rules address the needs and conditions of the community as a whole by regulating the development and uses of property. Zoning restrictions may apply to such various cases as whether a single-family house can be remodeled into apartments, whether a commercial space can be converted to residential use or the permissible height of a house or outbuilding.

It is advisable to apply to the local zoning board for approval before undertaking any kind of construction or remodeling that involves a house exterior or yard or if the project will substantially change the way a property is used. If the project does not conform with the standing zoning guidelines, you may apply to the zoning board for a variance. It is best to enlist the help of an architect or attorney for this.

Landmark regulations

Homes in historic districts may be subject to restrictions placed to help the neighborhood retain its architectural distinction and character. For the most part in designated landmark areas, changes in house exteriors are closely regulated. While extensive remodeling that would significantly change the architectural style are almost never permitted, even seemingly small modifications of existing structures are scrutinized. For example, metal or vinyl replacement windows may not be permitted for Victorian homes in designated areas, or the exterior paint and roof colors may be subject to approval. Even the color of the mortar used to repoint brickwork may be specified by the local landmarks commission or similar regulating body. Designs for new construction must conform to the prevalent architectural character. If you live in an historic district, it is advised that you apply to the governing body for approval of any plans for exterior renovation.

WILL YOU NEED A PERMIT OR VARIANCE?

Building code requirements and zoning regulations vary from city to city and frequently have county and state restrictions added to them. For this reason, it is impossible to state categorically which home-improvement projects require official permission and which do not. The chart below, which lists some of the most frequently undertaken projects, is meant to serve as a rough guide. Taken as a whole, it suggests a certain logic for anticipating when and what type of approval may be needed. Whether or not official approval is required, all work should be carried out in conformity with local code standards.

TYPE OF WORK	BUILDING PERMIT NEEDED		ZONING APPROVAL NEEDED	
Exterior painting and repairs / Interior decoration and repairs	NO	Permit or rigger's license may be needed to erect exterior scaffolding	NO	Certificate of appropriateness may be needed in historic areas
Replacing windows and doors	NO		NO	Permissible styles may be restricted in historic districts
Electrical	NO	Have work performed or checked by a licensed contractor	NO	Outdoor lighting may be subject to approval
Plumbing	NO	Have work performed or checked by a licensed contractor	NO	Work involving new water supply, septic or sewerage systems may require county health department
Heating	NO		NO	Installation of new oil storage tanks may require state environmental agency approval
Constructing patios and decks	Possibly		Possibly	
Installing a hot tub	NO		NO	
Structural alterations	YES		NO	Unless alterations change building height above limit or proximity of building to lot line
Attic remodeling	NO	Ascertain whether joists can safely support the floor load	NO	
Building a fence or garden wall	NO		YES	In cases where structure is adjacent to public road or easement or extends above a height set by board
Planting a hedge	NO		NO	
Path or sidewalk paving	NO		Possibly	Public sidewalks must conform to local standards and specifications
Clearing land	NO		YES	County and state environmental approval may also be needed
Constructing an in-ground pool	YES		YES	County and state environmental approval may also be needed
Constructing outbuildings	YES	For buildings larger than set limit	Possibly	
Adding a porch	NO	Unless larger than set limit	Possibly	Regulations often set permissible setback from public road
Adding a sunspace or greenhouse	YES		Possibly	Yes, if local rules apply to extensions
Constructing a garage	YES		Possibly	Yes, if used for a commercial vehicle and within set proximity to lot line
Driveway paving	NO		Possibly	Yes where access to public road created, also restrictions on proximity to lot lines
Constructing a house extension	YES		Possibly	Regulations may limit permissible house size and proximity to lot lines
Demolition	YES	If work involves structural elements	NO	Structures in historic districts may be protected by regulations
Converting 1-family house to multiunit dwelling	YES	Fire safety and ventilation codes are frequently more stringent for multiple dwellings	YES	
Converting a residential building to commercial use	YES		YES	

PLUMBER'S AND METALWORKER'S TOOL KIT

The growing use of plastics in plumbing is likely to affect the trade considerably, and while plastics have been used for drainage for some years, the advent of plastics suitable for mains, pressure and hot water lines will have the greatest impact. But brass fittings and pipework of copper and other metals are still the most commonly used for domestic plumbing, so the plumber's tool kit is still basically a metalworker's kit.

SINK- AND DRAIN-CLEANING EQUIPMENT

There's no need to hire a plumber to clear blocked appliances, pipes or even main drains. The necessary equipment can be bought or rented.

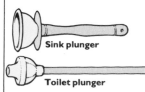

Sink plunger

Toilet plunger

Plunger
This is a simple but effective tool for clearing a blockage from the trap or toilet bowl. A pumping action on the rubber cap forces air and water along the pipe to disperse the blockage.

When you buy a plunger, make sure that the cup is big enough to surround the waste outlet completely. The cup of a toilet plunger may have a cone that makes a tight fit in the trap.

Compressed-air gun
A blocked wastepipe can be cleared with a compressed-air gun, if allowed by code. A hand-operated pump compresses air in the gun's reservoir, to be released into the pipe by a trigger. The gun has three interchangeable nozzles to suit different outlets.

Toilet auger
The short coiled-wire toilet auger, designed for clearing toilet traps, is rotated by a handle in a hollow rigid shaft. A vinyl guard prevents scratching of the toilet bowl.

• **Essential tools**
Sink plunger
Scriber
Center punch
Steel rule
Try square
General-purpose hacksaw

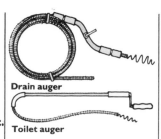

Drain auger

Toilet auger

Drain auger
A flexible drain auger of coiled wire will pass through small-diameter wastepipes to clear blockages. Pass the corkscrewlike head into the pipe until it reaches the blockage, clamp on the cranked handle and turn it to rotate the head and engage the blockage. Push and pull the auger until the pipe is clear.

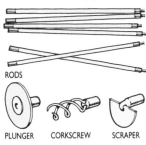

RODS

PLUNGER   CORKSCREW   SCRAPER

Drain rods
You can rent a complete set of rods and fittings for clearing main drains and inspection chambers. Traditionally, the rods are of flexible cane and wire, but modern ones come in lengths of polypropylene with threaded brass connectors. The clearing heads comprise a double-worm corkscrew fitting, small diameter rubber plunger and a hinged scraper for clearing the open channels in inspection chambers.

MEASURING AND MARKING TOOLS

Tools for measuring and marking metal are much like those used for wood but are made and calibrated for greater accuracy because metal parts must fit with great precision.

Scriber
For precise work, mark lines and hole centers on metal with a pointed hardened-steel scriber—but use a pencil to mark the center of a bend, as a scored line made with a scriber may open up on the outside of the bend when the metal is stretched.

Spring dividers
Spring dividers are like a pencil compass, but both legs have steel points. These are adjusted to the required spacing by a knurled nut on a threaded rod that links the legs.

Using spring dividers
Use dividers to step off divisions along a line (1) or to scribe circles (2). By running one point against the edge of a workpiece you can also scribe a line parallel with the edge (3).

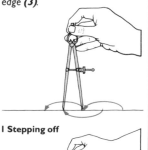

1 Stepping off

2 Scribing a circle

3 Parallel scribing

Center punch
A center punch is for marking the center of holes to be drilled.

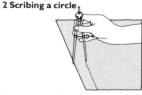

Correcting a center mark

Using a center punch
Place the punch's point on dead center and strike it with a hammer. If the mark is not accurate, angle the punch toward the true center, tap it to extend the mark in that direction, then mark the center again.

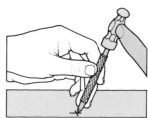

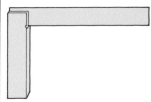

Steel rule
You'll need a long tape measure for estimating pipe runs and positioning appliances, but use a 1- or 2-foot steel rule for marking out components when absolute accuracy is important.

Try square
You can use a woodworker's try square to mark out or check right angles, but an all-metal engineer's try square is precision-made for metalwork. The small notch between blade and stock allows the tool to fit properly against a right-angled workpiece even when the corner is burred by filing. For general-purpose work, choose a 6-inch try square.

METAL-CUTTING TOOLS

You can cut solid bar, sheet and tubular metal with a hacksaw, but special tools for cutting sheet metal and pipes will give you more accuracy and speed the work.

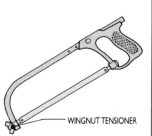

WINGNUT TENSIONER

General-purpose hacksaw
A modern hacksaw has a tubular steel frame with a light cast-metal handle. The frame is adjustable to blades of different lengths, which are tensioned by tightening a wingnut.

CHOOSING A HACKSAW BLADE

There are 8-, 10- and 12-inch hacksaw blades. Try the different lengths until you find which suits you best. Choose the hardness and size of teeth according to the metal to be cut.

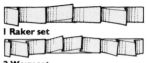

1 Raker set

2 Wavy set

Size and set of teeth
There are fine and coarse hacksaw blades, graded by the number of teeth per inch. A coarse blade has 14 to 18 teeth per inch and a fine one has 24 to 32. The teeth are set—bent sideways—to make a cut wider than the blade's thickness and prevent it from jamming in the work. Coarse teeth are "raker set" **(1)**, with pairs of teeth bent in opposite directions and every third or fifth one left in line with the blade to clear the cut of metal waste. Fine teeth are too small to be raker set and the whole row is "wavy set" **(2)**.

Use a coarse blade for cutting soft metals like brass and aluminum, which would clog fine teeth, and a fine blade for thin sheet and the harder metals.

Hardness
A hacksaw blade must be harder than the metal it is cutting or its teeth will quickly blunt. A flexible blade with hardened teeth will cut most metals, but there are fully hardened blades that stay sharp longer and are less prone to losing teeth. Being rigid and brittle, they break easily. High-speed steel blades for sawing very hard alloys are expensive and even more brittle than the fully hardened type.

Fitting a hacksaw blade
Adjust the length of the saw frame and slip the blade onto the pins at its ends, teeth pointing away from the handle. Then apply tension with the wingnut. If the new blade wanders off line when you work, tighten the wingnut.

If you have to fit a new blade after starting to cut a piece of metal it may jam in the cut because its set is wider than that of the old worn blade. Start a fresh cut on the other side of the workpiece and work back to the cut you began with.

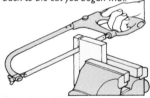

Turning a blade
Sometimes it's easier to work with the blade at right angles to the frame. Rotate the blade-attachment bars a quarter turn before fitting the blade on the pins.

1 Turn first kerf away from you

Sawing metal bar
*Hold the work in a machinist's vise, the marked cutting line as close to the jaws as possible. Start the cut on the waste side of the line with short strokes until it is about 1/16 inch deep, then turn the bar 90 degrees in the vise, so that the cut faces away from you, and make a similar one in the new face **(1)**. Continue in this way until the kerf (cut) runs right around the bar, then cut through the remainder with long steady strokes. Steady the end of the saw with your free hand and put a little light oil on the blade if necessary.*

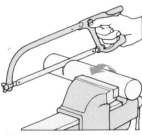

Sawing rod or pipe
As you cut a cylindrical rod or tube with a hacksaw, rotate the work away from you until the kerf goes right around it, then cut through it.

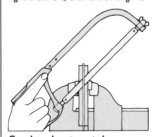

Sawing sheet metal
To saw a small piece of sheet metal, sandwich it between two strips of wood clamped in a vise. Adjust the metal to place the cutting line close to the strips, then saw down the waste side with steady strokes and the blade angled to the work.
Clamp thin sheet metal between two pieces of plywood and cut through all three layers at once.

Sawing a groove
To cut a slot or groove wider than a standard hacksaw blade, fit two or more identical blades in the frame at the same time.

Junior hacksaw
Use a junior hacksaw for cutting small-bore tubing and thin metal rod. In most types the blade is held under tension by the spring steel frame.

Fitting a new blade
To fit a blade, locate it in the slot at the front of the frame, then bow the frame against a workbench until the blade fits in the rear slot.

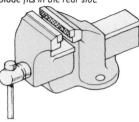

Machinist's vise
A large machinist's or metalworker's vise is bolted to the workbench, but small clamp-on ones are also available. Slip soft fiber liners over the jaws of a vise to protect workpieces held in it.

Cold chisel
Though plumbers use cold chisels to hack old pipework out of masonry, they are also for cutting metal rod and slicing the heads off rivets. Keep the tip of yours sharpened on a bench grinder.

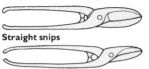

Straight snips

Universal snips

Tinsnips
Tinsnips are heavy-duty scissors for cutting sheet metal. Straight snips have wide blades for cutting straight edges, and if you try to cut curves with them the waste gets caught against the blades. But it is possible to cut a convex curve by removing small straight pieces of waste and working down to the marked line. Universal snips have thick, narrow blades that will cut a curve in one pass and will also make straight cuts.

Using tinsnips
As you cut along the marked line, let the waste curl away below the sheet. If the metal is too thick to be cut with one hand, clamp one handle of the snips in a vise so that you can put your full weight on the other.

Try not to close the jaws completely each time, as it can cause a jagged edge on the metal. Wear thick work gloves when you are cutting sheet metal.

SHARPENING SNIPS

Clamp one handle in a vise and sharpen the edge with a smooth file; repeat with the other and finish by removing the burrs from the backs of the blades on an oiled slipstone.

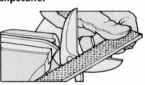

Sheet-metal cutter
Unlike tinsnips, which will distort a narrow waste strip on one side of the cutting line, a sheet-metal cutter removes a narrow strip as perfectly flat as the larger sheet. It is also suited to cutting rigid plastic sheet, which can crack if it is distorted by tinsnips.

Tube cutter
A tube cutter will cut the ends of pipes at exactly 90 degrees to their length. The pipe is clamped between the cutting wheel and an adjustable slide with two rollers, and is cut as the tool is revolved around it and the adjusting screw tightened before each turn. Keep the cutter lightly oiled when you use it.

Chain-link cutter
Cut large-diameter pipes with a chain-link cutter. Wrap the chain round the pipe, locate the end link in the clamp and tighten the adjuster until the cutter on each link bites into the metal. Work the handle back and forth to score the pipe and continue tightening the adjuster intermittently to cut deeper, until the pipe is severed.

SEE ALSO

Details for: ▷

| Cutting pipe | 9, 11 |

Sheet-metal cutter

Tube cutter

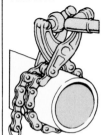

Chain-link cutter

- **Essential tools**
  Junior hacksaw
  Cold chisel
  Tinsnips
  Pipe cutter

DRILLS AND PUNCHES

Special-quality steel bits are made for drilling holes in metal. Cut 1/2- to 1-inch holes in sheet metal with a punch.

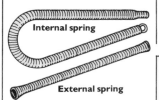

Twist drills

Metal-cutting twist drills are much like those used for wood but are made from high-speed steel and with tips ground to a shallower angle. Use them in a power drill at slow speeds.

Mark the hole's center with a center punch to locate the drill point and clamp the work in a vise or to the bed of a drill stand. Drill slowly and steadily and keep the bit oiled.

To drill a large hole, make a pilot hole first with a small drill to guide the larger one.

When drilling sheet metal, the bit may jam and produce a ragged hole as it exits on the far side of the work. To prevent this, clamp the work between pieces of plywood and drill through the three layers.

Hole punch

Use a hole punch to make large holes in sheet metal. Mark the circumference of the hole with spring dividers, lay the metal on a piece of scrap softwood or plywood, place the punch's tip over the marked circle, tap it with a hammer, then check the alignment of the punched ring with the scribed circle. Reposition the punch and, with one sharp hammer blow, cut through the metal. If the wood gives and the metal is slightly distorted, tap it flat again with the hammer.

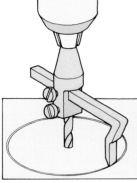

Tank cutter

Use a tank cutter to make holes for pipework in plastic or metal storage cisterns.

METAL BENDERS

Thick or hard metal must be heated before it can be bent successfully, but soft copper piping and sheet metal can be bent while cold.

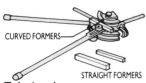

Internal spring

External spring

Bending springs

You can bend small-diameter pipes over your knee, but their walls must be supported with a coiled spring to prevent them from buckling.

Push an internal-type spring inside the pipe or an external-type one over it. Either type of spring must fit the pipe exactly.

CURVED FORMERS

STRAIGHT FORMERS

Tube bender

In a tube bender, pipe is bent over one of two fixed curved formers that give the optimum radii for plumbing. Each has a matching straight former which is placed between the pipe and a steel roller on a movable lever. When this lever is moved towards the fixed one, the pipe is bent over the curved former. The formers support the pipe walls during bending.

You can get extra leverage by clamping the fixed lever in a vise and using both hands on the movable one.

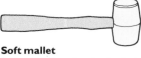

Soft mallet

Soft mallets have heads of coiled rawhide, hard rubber or plastic. They are used in bending strip or sheet metal, which would be damaged by a metal hammer.

To bend sheet metal at a right angle, clamp it between stout wood strips along the bending line. Start at one end and bend the metal over one of the strips by tapping it with the mallet. Don't attempt the full bend at once but work along the sheet, increasing the angle and keeping it constant along the length until the metal lies flat on the strip. Then knock out any kinks.

TOOLS FOR JOINING METAL

You can make permanent watertight joints between metal components by using a molten alloy that acts like a glue when it cools and solidifies. Mechanical fasteners like compression joints, rivets and nuts and bolts are also used for joining metal.

SOLDERS

Solders are special alloys for joining metals and are designed to melt at temperatures lower than the melting points of the metals to be joined. Soft solder, a tin-and-lead alloy, melts at 183° to 250°C. Brazing, a method of hard soldering involving a copper-and-zinc alloy, requires the even higher temperature of some 850° to 1000°C.

Solder is available as a coiled wire or a thick rod. Use soft solder for copper plumbing fittings.

FLUX

To be soldered, a joint must be perfectly clean and free of oxides. Even after cleaning with steel wool or emery cloth, oxides will form immediately, preventing a positive bond between solder and metal. Flux forms a chemical barrier against oxidation. Corrosive, or "active" flux, applied with a brush, actually dissolves oxides but must be washed from the surface with water as soon as the solder solidifies or it will go on corroding the metal. A "passive" flux, in paste form, is used where it will be impossible to wash the work thoroughly. Though a passive flux will not dissolve oxide, it will exclude it adequately for soldering copper plumbing joints and electrical connections.

Some wire solder contains flux in its hollow core. The flux flows before the solder melts.

Soldering irons

Successful soldering needs the work to be made hot enough to melt the solder and cause it to flow; otherwise the solder will solidify before it can completely penetrate the joint. The necessary heat is applied with a soldering iron.

Pencil-point iron

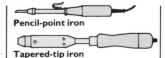

Tapered-tip iron

There are simple irons that are heated in a fire, but an electric iron is much handier to use and its temperature is both controllable and constant. Use a low-powered pencil-point iron for soldering electrical connections and a larger one with a tapered tip to bring sheet metal up to working temperature.

Tinning a soldering iron

The tip of a soldering iron must be coated with solder to keep it oxide-free and maintain its performance. Clean the cool tip with a file, then heat it to working temperature, dip it in flux and apply a stick of solder to coat it evenly.

Using a soldering iron

Clean the mating surfaces of the joint to a bright finish and coat them with flux, then clamp the joint tightly between two wooden strips. Apply the hot iron along the joint to heat the metal thoroughly, then run its tip along the edge of the joint, following it with solder. Solder will flow immediately into a properly heated joint.

Gas torch

Even a large soldering iron cannot heat thick metal fast enough to compensate for heat loss away from the joint, and this is very much the situation when you solder pipework. Though the copper unions have very thin walls, the pipe on each side dissipates so much heat that an iron cannot get the joint itself hot enough to form a watertight soldered seal. Use a gas torch with an intensely hot flame that will heat the work quickly.

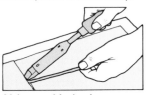

- **Essential tools**
  High-speed twist drills
  Variable or two-speed power drill
  Bending spring
  Soft mallet
  Soldering iron
  Gas torch

The torch runs on propane or Mapp contained under pressure in a disposable metal canister that screws onto the gas inlet. Open the control valve and light the gas released from the nozzle, then adjust the valve until the flame roars and is bright blue. Use the hottest part of the flame—about the middle of its length—to heat the joint.

Fiberglass mat
Buy a fireproof mat of fiberglass from a plumbers' supply to protect flammable surfaces from the heat of a gas torch.

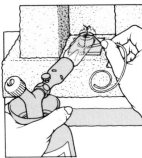

Hard-soldering and brazing
Use a gas torch to hard-solder or braze. Clean and flux the work—if possible with an active flux—then wire or clamp the parts together. Place the assembly on a fireproof mat or surround it with firebricks. Bring the joint to red heat with the torch, then dip a stick of the appropriate alloy in flux and apply it to the joint. When the joint is cool, chip off hardened flux, wash the metal thoroughly in hot water and finish the joint with a file.

Hot-air gun
Some hot-air guns, designed for stripping old paintwork, can also be used for soft soldering. You can vary the temperature of an electronic gun from 100° to 600°C. A heat shield on the nozzle reflects the heat back onto the work.

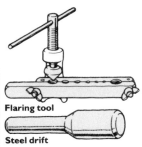

Flaring tool

Steel drift

Flaring tools
To make a nonmanipulative joint in plumbing pipework, the ends of the copper pipes are simply cut square and cleaned up to remove sharp edges. For a manipulative joint, the pipe ends must be flared to lock into the joint.

The simplest way to flare a copper pipe is with a steel drift. First slip one joint capnut onto the pipe, then push the narrow shank of the drift into the pipe. Strike the tool with a heavy hammer to drive its conical part into the pipe, stretching the walls to the required shape.

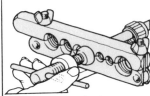

Using a flaring tool
A flaring tool will shape pipework more accurately. Again with the capnut in place, clamp the pipe in the matching hole in the die block, its end flush with the side of the block, then turn the screw of the flaring tool to drive its cone into the pipe.

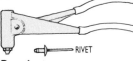

Pop riveter
Join thin sheet metal with a pop riveter, a hand-operated tool with pliers-like handles. The special rivets have long shanks that break off and leave slightly raised heads on both sides of the work.

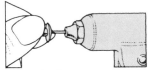

1 Insert the rivet

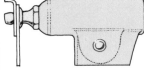

2 Squeeze the handles

Using a riveter
Clamp the two sheets together and drill holes right through the metal, matching the diameter of the rivets and spacing them regularly along the joint. Open the handles of the riveter and insert the rivet shank in the head (1). Push the rivet through a hole in the work and, while pressing the tool hard against the metal, squeeze the handles to compress the rivet head on the far side (2). When the rivet is fully expanded, the shank will snap off in the tool.

WRENCHES

A plumber uses a great variety of wrenches on a wide range of fittings and fixings, but you can rent the ones that you need only occasionally.

Open-end wrench
A set of the familiar open-end wrenches is essential to a plumber or metalworker. In most situations, pipework runs into a fitting or accessory and it is not possible to use anything but a wrench with open jaws. Most wrenches are double-ended, perhaps in a combination of sizes, and sizes are duplicated within a set for when two identical nuts have to be turned simultaneously, as on a compression joint, for instance.

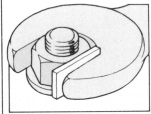

Achieving a tight fit
A wrench must be a good fit or it will round over the corners of the nut. You can pack out the jaws with a thin "shim" of metal if a snug fit is otherwise not possible.

Box wrench
Being a closed circle, the head of a box wrench is stronger and fits better than that of an open-ended one. It is especially handy for loosening corroded nuts if it can be slipped onto them from above.

Square nut　　**Hexagonal nut**

Choosing a box wrench
Choose a 12-point wrench. It is fast to use and will fit both square and hexagonal nuts. You can buy combination wrenches with rings at one end and open jaws at the other.

Socket tube
A socket tube is a steel tube with its ends shaped into hexagons— an excellent tool for reaching a nut in a confined space. Turning force is applied with a bar slipped into a hole drilled through the tube. Don't use a very long bar. Too much leverage may strip the thread of the fitting or distort the thin walls of the tube.

Adjustable wrench
Having a movable jaw, an adjustable wrench is not as strong as an open-end or box wrench but is often the only tool that will fit a large or painted-over nut. Make sure the wrench fits the nut snugly by rocking it slightly as you tighten the jaws. Grip the nut with the roots of the jaws. If you use just the tips, they can spring apart slightly under force and the wrench will slip.

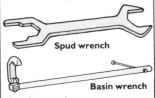

Spud wrench

Basin wrench

Spud wrench and basin wrench
A spud wrench is a special double-ended wrench for use on large toilet and sink nuts.

A basin wrench has a pivoting jaw that can be set for either tightening or loosening hard-to-reach fittings.

Radiator wrench
Use a simple Allen wrench of hexagonal-section steel rod to remove radiator drain plugs. One end is ground to fit plugs with square sockets.

● **Essential tools**
Blind riveter
Set of open-ended wrenches
Small and large adjustable wrenches

SEE ALSO

◁ Details for:
Working with pipe   8–12

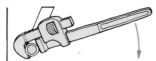

Pipe wrench
The adjustable toothed jaws of a pipe wrench are for gripping pipework. As force is applied, the jaws tighten on the work.

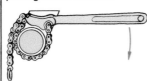

Chain wrench
A chain wrench does the same job as a pipe wrench but it can be used on very large-diameter pipes and fittings. Wrap the chain tightly around the work and engage it with the hook at the end of the wrench, then lever the handle towards the toothed jaw to apply turning force.

Strap wrench
With a strap wrench you can disconnect chrome pipework without damaging its surface. Wrap the smooth leather or canvas strap around the pipe, pass its end through the slot in the head of the tool and pull it tight. Leverage on the handle will rotate the pipe.

Locking pliers
These special pliers lock onto the work. They will grip round stock or damaged nuts and are often used as a small clamp.

Using locking pliers
Squeeze the handles to close the jaws while slowly turning the adjusting screw clockwise until they snap together (1). Release the tool's grip on the work by pulling the release lever (2).

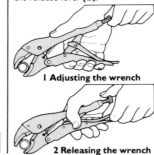

1 Adjusting the wrench

2 Releasing the wrench

- **Essential tools**
  Locking pliers
  Second-cut and
  smooth flat files
  Second-cut and
  smooth half-round
  files

FILES
Files are used for shaping and smoothing metal components and removing sharp edges.

CLASSIFYING FILES
The working faces of a file are composed of parallel ridges, or teeth, set at about 70 degrees to its edges. A file is classified according to the size and spacing of its teeth and whether it has one or two sets.

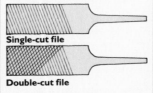

Single-cut file

Double-cut file

A *single-cut file* has one set of teeth virtually covering each face. A *double-cut file* has a second set of identical teeth crossing the first at a 45-degree angle. Some files are single-cut on one side and double-cut on the other.

The spacing of the teeth relates directly to their size: The finer the teeth, the more closely packed they are. Degrees of coarseness are expressed in numbers of teeth per inch. Use progressively finer files to shape a component and to gradually remove marks left by coarser ones.

File classification
Bastard file—Coarse grade (26 teeth per inch)—For initial shaping. ***Second-cut file***—Medium grade (36 teeth per inch)—For preliminary smoothing. ***Smooth file***—Fine grade (47 teeth per inch)—For final smoothing.

CLEANING A FILE
Soft metal clogs the teeth of files but can be removed by brushing along the teeth with a file card—a fine wire brush. Chalk rubbed on a clean file will reduce clogging.

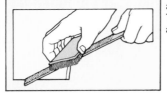

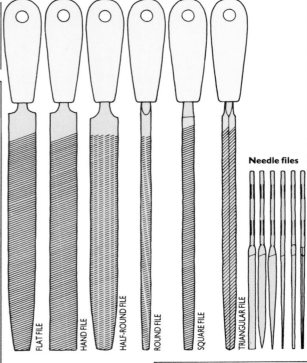

FLAT FILE · HAND FILE · HALF-ROUND FILE · ROUND FILE · SQUARE FILE · TRIANGULAR FILE

Needle files

Flat file
A flat file tapers from its pointed tang to its tip, both in width and thickness. Both faces and both edges are toothed.

Hand file
Hand files are parallel-sided but tapered in their thickness. Most of them have one smooth edge for filing up to a corner without damaging it.

Half-round file
This tool has one rounded face for shaping inside curves.

Round file
A round file is for shaping tight curves and enlarging holes.

Square file
This file is for cutting narrow slots and smoothing the edges of small rectangular holes.

Triangular file
The triangular file is for accurately shaping and smoothing undercut apertures of less than 90 degrees.

Needle files
Needle files are miniature versions of standard files and are all made in extra-fine grades. They are used for precise work and to sharpen brace bits.

FILE SAFETY
Always fit a wooden or plastic handle on the tang of a file before using it.

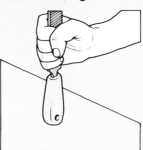

1 Fitting a file handle

2 Knock a handle from the tang

If an unprotected file catches on the work, the tang could be driven into the palm of your hand. Having fitted a handle, tap its end on a bench to tighten its grip (**1**).

To remove a handle, hold the file in one hand and strike the ferrule away from you with a piece of wood (**2**).

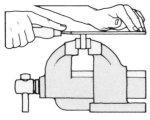

Using a file
When you use any file, keep it flat on the work and avoid rocking it on the forward strokes. Hold it steady with the fingertips of one hand on its tip and make slow, firm strokes with the full length of the tool.

To avoid vibration, hold the work low in a vice or clamp it between two strips of wood.

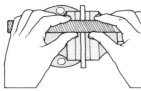

Draw filing
You can give metal a smooth finish by draw filing. With both hands, hold a smooth file at right angles to the work and slide it back and forth along the surface, then polish the work with emery cloth wrapped around the file.

PLIERS

Pliers are for improving your grip on small components and for bending and shaping metal rod and wire.

Machinist's pliers
For general-purpose work, buy a sturdy pair of machinist's pliers. The toothed jaws have a curved section for gripping round stock and side cutters for cropping wire.

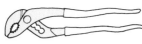

Slip-joint
or waterpump pliers
The special feature of slip-joint pliers is a movable pivot for enlarging the jaw spacing. The extra-long handles give a good grip on pipes and other fittings.

FINISHING METAL

Apart from its appearance, metal must be clean and rust-free if it is to be painted or soldered.

Wire brush
Use a steel wire hand brush to clean rusty or corroded metal.

Steel wool
Steel wool is a mass of very thin steel filaments. It is used to remove file marks and to clean oxides and dirt from metals.

Emery cloth and paper
Emery is a natural black grit. Backed with paper or cloth for polishing metals, it is available in a range of grades from coarse to fine. For the best finish, work through the grades, using progressively finer abrasives as the work proceeds.

1 Glue paper to a board

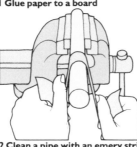

2 Clean a pipe with an emery strip

Using emery cloth and paper
To avoid rounding the crisp edges of a flat component, glue a sheet of emery paper to a board and rub the metal on it (*1*).

To finish round stock or pipes, loop a strip of emery cloth over the work and pull alternately on each end (*2*).

Buffing wheel
Metals can be brought to a shine by hand using liquid polish and a soft cloth, but for a really high gloss, use a buffing wheel in a bench-mounted power drill or grinder.

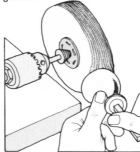

Using a buffing wheel
Apply a stick of buffing compound—a fine abrasive with wax—to the revolving wheel, then move the work from side to side against its lower half, keeping any edges facing down.

Reseating tool
When the seat of a faucet is so worn that even a new washer will not make a perfect seal, you can grind a new seat with a reseating tool. Remove the valve parts and screw the cone into the body of the faucet. Lower the threaded section with the knurled adjuster, then turn the bar to operate the cutter.

WOODWORKING TOOLS

A plumber needs basic woodworking tools to lift floorboards, notch joists for pipe runs and attach pipe clips.

• **Essential tools and materials**
Machinist's pliers
Wire brush
Steel wool
Emery cloth and paper

GLOSSARY OF TERMS

ABS pipe
A plastic drainage pipe made from acrylonitrile butadiene styrene; may be used for drainage where permitted.

Air filter
Traps airborne particles in a forced-air system.

Airlock
A blockage in a pipe caused by a trapped bubble of air.

Appliance
A machine or device powered by electricity. Also, a functional piece of equipment connected to the plumbing—a basin, sink, bath, etc.

Back-siphonage
The siphoning part of a plumbing system caused by the failure of main water pressure.

Balanced flue
A ducting system which allows a heating appliance, such as a boiler, to draw fresh air from, and discharge gases to, the outside of a building.

Bleed valve
A valve on one side of a radiator that is loosened to purge air.

Blower
The part of a forced-air system that circulates air.

Burner
The component on a furnace that heats air or water and that burns oil or gas to produce the heat.

Burr
The rough raised edge left on a pipe after cutting or filing.

Cesspool
A covered or buried tank for the collection and storage of sewage.

Circulator pump
A motorized device that controls the flow of water to the radiators or convectors in a hot-water system.

Cleanout
An opening near the bottom of the stack in a home's plumbing system that can be accessed and cleaned if drainage is sluggish.

Convector
A heater that emits none of its own heat in the form of direct radiation; hot water passes through a finned pipe inside the heater and the fins absorb the heat and transfer it to the air around them.

CPVC pipe
A type of plastic piping made from chlorinated polyvinyl chloride.

Damper
A paddlelike device inside a duct that regulates airflow to different parts of a house in a forced-air system.

Dielectric union
A special connector used when copper pipe is joined to steel pipe. It prevents an electrolytic reaction that causes corrosion when these dissimilar metals are joined together.

Draincock
A tap from which a plumbing system or appliance is drained.

Drain valve
A spigot at the base of a water heater that releases water when opened.

Duct
A metal passageway that carries air to and from the furnace in a forced-air system.

Expansion tank
A cylinder attached to a boiler that relieves excess system pressure that is created when water in a hot-water heating system heats and expands.

Filter
An attachment on the water supply pipe or tap that screens out sediment, minerals, and chemicals from drinking water.

Fixture
A bathtub, shower, sink, or toilet.

Flashing
A weatherproof metal junction between the roof and the top of the stack in the plumbing system.

Flue liner
A lining in a chimney that protects the brickwork from the corrosive elements in rising smoke.

Fuse box
The point where the main electrical service cable is connected to the house circuitry in an older home. *See also* service panel.

Galvanized pipe
A pipe covered with a protective coating of zinc.

Gate valve
A valve on both sides of a water meter that controls the flow of water into a home from the main supply coming from the utility company or municipality.

Ground-wire jumper
A continuous ground wire that must be run between two metal sections of pipe that have plastic piping between them when a house's electrical system is grounded to the plumbing.

Hood
A metal overhang installed across a fireplace opening that traps smoke.

Insulation
Materials used to reduce the transmission of heat or sound. Also a nonconductive material surrounding electrical wires or connections to prevent the passage of electricity.

Lead
A component in solder used to join runs of copper pipe. Solder with a high lead content has the potential to cause lead poisoning.

Overflow pipe
A drainage pipe designed to safely discharge water that has risen above its intended level in a water heater or boiler.

PB pipe
A type of flexible plastic pipe made from polybutylene which may be used where permitted by local plumbing codes.

Pressure relief valve
A safety feature that releases excess pressure that builds up inside a boiler or hot-water heater.

PVC pipe
A type of plastic drainage pipe made from polyvinyl chloride which may be used where permitted by local plumbing codes.

Rust inhibitor
A chemical added to water in a boiler that prevents corrosion in the pipes of a hot-water system.

Seepage pit
A miniature leach field designed to dispose of gray water discharged from appliances and floor drains.

Septic tank
A sewage-storage tank, similar to a cesspool, but the waste is treated to render it harmless before it is discharged into a local waterway or under ground.

Service panel
The point where the main electrical service cable is connected to the house circuits. Circuit breakers protect individual circuits in the system.

Service pipe
The supply pipe bringing water into a house and which is connected to the rising main.

Shutoff valve
A valve connecting water to plumbing fixtures; allows user to shut off water to an individual fixture.

Sillcock
A stem-and-seat faucet with a long stem typically installed as an outdoor faucet. Freezeless sillcocks may be left on during freezing weather.

Softener
A machine that treats mineral salts in water, which can build up and cause clogs in pipes and give water an unpleasant odor.

Teflon tape
Material that is wrapped around the threads of pipe fittings to prevent water and gas leaks.

Thermostat
A device that maintains a heating system at a constant temperature.

Timer
A device that can be set to switch a heating system on and off to suit the heating needs of a home's occupants.

Trap
A bent section of pipe containing standing water to prevent the passage of gases; sometimes called a P-trap.

Vent
A device that permits air to bleed out automatically as a steam radiator heats up.

Volt-ohmmeter
A device that can measure resistance and voltage.

Water hammer
A vibration in plumbing pipework produced by fluctuating water pressure.

Water-resistant wallboard
Wall panels made for use in damp locations like bathrooms and kitchens.

Zone control valve
A motorized valve that directs the flow of heating water from a boiler through preselected pipes at preselected times.